Quantum Phenomena in Magnetic Resonance

Exploring the Fundamental Principles of Nuclear and Electron Spins

Drew Fitzpatrick

Table of Contents

1 Introduction to Quantum Mechanics

The cornerstone of quantum mechanics lies in the radical departure from classical physics. Forget the deterministic, predictable world of Newtonian mechanics; instead, embrace probability and uncertainty. At the heart of this lies the quantization of energy. Unlike classical systems where energy can take on any value, quantum systems are restricted to discrete energy levels. Imagine a staircase instead of a ramp; you can only stand on specific steps, not anywhere in between. This seemingly simple concept has profound implications, shaping the behavior of atoms, molecules, and ultimately, the universe itself. Understanding this fundamental principle is vital for grasping the essence of quantum phenomena in magnetic resonance.

Wave-particle duality is another key concept that challenges our classical intuition. Light, previously considered a wave, also exhibits particle-like properties. Conversely, particles, like electrons, demonstrate wave-like behavior. This dual nature isn't a simple either/or; it's a fundamental property of quantum objects. They exist as both wave and particle simultaneously, a superposition of states, a concept we'll explore more deeply in the next section. This duality is central to the workings of NMR and EPR, affecting how particles interact with magnetic fields and how we measure them.

The uncertainty principle, proposed by Heisenberg, states that there's a fundamental limit to the precision with which certain pairs of physical properties of a particle, like position and momentum, can be known simultaneously. The more accurately you know one, the less accurately you know the other. This isn't a matter of measurement limitations; it's a fundamental property of the quantum world. It implies inherent fuzziness and unpredictability at the atomic level, which significantly impacts the interpretation of experimental data in quantum resonance studies. This principle directly challenges our classical understanding of deterministic systems, requiring a probabilistic interpretation of quantum phenomena.

Quantum superposition, a direct consequence of wave-particle duality, is crucial for understanding many quantum effects. Unlike classical objects which exist in a definite state, a quantum system can exist in multiple states at once. A classic example is Schrödinger's cat, famously (and conceptually) both alive and dead simultaneously until observed. In the realm of magnetic resonance, atomic nuclei or electrons can exist in a superposition of spin states, allowing for coherent manipulation of their spin orientation through precisely controlled radiofrequency pulses. This principle forms the very foundation of NMR and EPR techniques.

The act of measurement itself is crucial. In classical physics, observation doesn't affect the system being measured. However, in the quantum world, measurement fundamentally alters the system's state. The very act of observing a quantum system collapses its superposition into a single definite state. This collapse is not a gradual process; it's abrupt and probabilistic, influenced by the nature of the measurement itself. Understanding this crucial aspect is fundamental to interpreting the outcomes of experiments in NMR and EPR, where the measurements affect the system's quantum state and hence its observable properties. This highlights the crucial interaction between the observer and the observed, a cornerstone of quantum mechanics.

Moving beyond these fundamental principles, we must introduce the concept of wave functions. These mathematical functions describe the quantum state of a system and contain all the information about its possible properties. The square of the wave function's magnitude gives the probability of finding the system in a particular state. This probabilistic interpretation is central to quantum mechanics. We don't know precisely where a particle is; we know the probability of it being in a particular location. This probabilistic nature of quantum mechanics underpins the interpretation of experimental data in magnetic resonance, where signals reflect probability distributions rather than deterministic outcomes. Analyzing these wave functions is crucial for understanding and predicting the behavior of quantum systems in magnetic resonance experiments.

Finally, understanding the mathematical framework underlying quantum mechanics is essential for a deeper appreciation of NMR and EPR. Operators, eigenstates, eigenvalues, and the time-dependent Schrödinger equation are fundamental mathematical tools used to describe the evolution of quantum systems and predict their behavior under various conditions. While not covered extensively here, familiarity with these concepts is necessary for a comprehensive understanding of the underlying physics of quantum resonance techniques. Mastering these mathematical tools enables you to build upon the fundamental concepts discussed, providing a strong foundation for deeper explorations into advanced NMR and EPR techniques. The applications are vast, spanning diverse fields, and understanding these fundamentals is the essential first step.

1.1 Basic Quantum Principles

The fundamental concept underpinning quantum mechanics is quantization. Unlike classical physics where energy, momentum, and other properties can take on any value within a continuous range, quantum mechanics dictates that these properties are often restricted to discrete, specific values. Imagine a staircase; you can only stand on specific steps, not between them. Similarly, an electron in an atom can only occupy specific energy levels, not intermediate ones. This quantization is crucial for understanding the behavior of nuclei and electrons in magnetic fields, the very foundation of NMR and EPR. These discrete energy levels are directly tied to the inherent angular momentum, or spin, of these particles, a property that has no classical analogue.

Energy quantization leads directly to the concept of energy transitions. An electron or nucleus can absorb a photon of precisely the right energy to jump to a higher energy level. Conversely, it can emit a photon and transition to a lower energy level. The energy difference between these levels directly determines the frequency of the photon involved. This precise energy matching, the fundamental principle behind resonance, is essential to both NMR and EPR. It's this resonance that allows us to probe the properties of matter at a microscopic level. We exploit this principle to investigate the inner workings of molecules and materials. This understanding is paramount to interpreting the spectra and extracting meaningful information from NMR and EPR experiments.

Unlike classical mechanics, which precisely predicts the trajectory of a particle given its initial conditions, the quantum mechanical world is probabilistic. We can only talk about the probability of finding a particle in a specific state or location. This is encapsulated in the wave function, a mathematical description of the quantum system that contains all the information we can know about it. The square of the wave function's magnitude gives us the probability density—the likelihood of finding the particle in a particular region of space. This inherent uncertainty is not due to limitations in our measurement apparatus but is a fundamental property of the quantum world. This probabilistic nature is critical for understanding the behavior of spins in NMR and EPR, as we are often dealing with ensembles of particles, each following its own probability distribution.

The wave-particle duality is another cornerstone of quantum mechanics. This principle states that quantum objects, such as electrons and photons, exhibit properties of both waves and particles. This duality is not merely a convenient way to describe their behavior, but reflects an inherent property of quantum systems. The wave-like nature manifests in phenomena such as diffraction and interference, while the particle-like nature is evident in interactions such as absorption and emission of photons. Understanding this dual nature is vital, because the signals we observe in NMR and EPR arise from interactions that fundamentally hinge upon both the wave and particle aspects of the quantum world. Consider the spin of a nucleus or electron; its interaction with a magnetic field can be described as a precession, a wave-like phenomenon, while the energy transitions involve the absorption or emission of photons - distinctly particle-like events.

Finally, let's briefly discuss the role of operators in quantum mechanics. Unlike classical mechanics where variables represent specific values, in quantum mechanics they are represented by operators. These operators act on the wave function to extract information about the physical quantities, like energy or momentum. The eigenvalue equation, where an operator acting on a wave function yields the wave function multiplied by a constant (the eigenvalue), is central to quantum mechanics. The eigenvalues then correspond to the possible measured values of the physical quantity. In NMR and EPR, this has direct application in understanding the energy levels and transition probabilities of nuclear and electron spins within a magnetic field. These operators and their eigenvalues form the mathematical framework for predicting and interpreting experimental data.

1.2 Quantum Superposition

Imagine a coin spinning in the air; it's neither heads nor tails until it lands. Quantum superposition is similar, but far stranger. Instead of a simple binary outcome, a quantum system—like an electron's spin—exists in a probabilistic blend of all possible states simultaneously. It's not just that we don't know its state; it genuinely is in multiple states at once, a bizarre reality defying classical intuition. This inherent uncertainty is key to understanding NMR and EPR.

This probabilistic nature is described mathematically using wave functions, complex mathematical descriptions encompassing all potential states. The probabilities of finding the system in a specific state are encoded within the wave function's squared magnitude. Measurement, the act of observing the system, forces a "collapse" of the wave function, selecting one specific state from the superposition. Before measurement, however, the system remains in its ambiguous, superimposed state.

Think of a spinning electron, capable of spinning "up" or "spinning down". Classically, it would be one or the other. However, in the quantum realm, it exists as a combination of both until measured. This superposition isn't simply a lack of knowledge; it's a fundamental property of the quantum world, where particles can simultaneously possess seemingly contradictory properties. This fundamental concept directly underpins the principles of NMR and EPR, as we will see later in the book.

The mathematical representation of superposition often involves linear combinations of basis states. For a spin-1/2 particle like an electron, the basis states are typically $|\uparrow\rangle$ (spin up) and $|\downarrow\rangle$ (spin down). A superposition state can then be expressed as $|\uparrow\rangle + |\downarrow\rangle$, where and are complex numbers representing the probability amplitudes of the spin being up or down respectively. The square of the absolute value of each amplitude determines the probability of obtaining that specific outcome upon measurement. This is a crucial concept; remember this mathematical formalism because it applies directly to the phenomena we will discuss later, regarding the specific spin states of atoms and molecules under magnetic resonance.

Superposition isn't just a theoretical curiosity; it has profound practical implications. In NMR, the application of radiofrequency pulses creates a superposition of nuclear spin states, initiating the process of resonance. Similarly, in EPR, microwave pulses manipulate electron spins, creating superpositions essential for observing the resonance signal. Without the possibility of these quantum superpositions, the very basis of NMR and EPR spectroscopy would collapse. These techniques depend on the ability to manipulate these probabilistic quantum states.

The lifetime of a superposition is limited by various factors including the interaction with the environment. These interactions cause decoherence, a process where the quantum system loses its coherence and the superposition collapses, transitioning into a definite state. Therefore, understanding and mitigating decoherence is a crucial aspect of advanced NMR and EPR techniques, particularly in quantum computing applications. The longer we can maintain the superposition, the more we can leverage its power for applications ranging from advanced medical imaging to designing advanced materials.

Understanding superposition, then, is not just about grasping a theoretical concept; it's about understanding the very foundation of NMR and EPR. It's the key that unlocks the door to manipulating the quantum world, allowing us to probe the intricate details of matter at the atomic and molecular levels. Mastering this concept will enable you to truly appreciate the power and potential of these remarkable spectroscopic techniques. We will use this foundation as a springboard to explore the more complex techniques and applications discussed in later chapters, bridging the gap between abstract quantum theory and the practical techniques used in modern scientific research.

The fragility of superposition highlights the delicate balance between the quantum world and the macroscopic world we experience. Interactions with the environment can swiftly destroy the superposition, converting it into a classical probability distribution. However, this does not diminish the importance of superposition; rather, it emphasizes the need for carefully controlled experimental conditions to observe and exploit quantum phenomena in NMR and EPR. The future of quantum technologies, including quantum computing and advanced sensing, hinges on our ability to control and manipulate these fragile states, extending their lifetimes and preventing decoherence.

This inherent instability underscores why studying superposition in the context of magnetic resonance is so vital. The magnetic field itself can influence the lifetime of a superposition. Careful control of the magnetic field parameters is crucial for maximizing signal strength and minimizing experimental artifacts in both NMR and EPR. By understanding the factors that influence superposition's lifetime, researchers can optimize their experiments, improving the accuracy and resolution of their measurements.

The transition from the concept of superposition to the practical application of NMR and EPR involves a careful consideration of experimental parameters, pulse sequences, and data analysis techniques. While the theoretical concept of superposition is fundamental, the reality of working with quantum systems demands a deep understanding of the experimental realities and the limitations imposed by decoherence. Understanding these practical aspects is just as crucial as understanding the underlying theory.

In the following chapters, we will delve into the specific applications of superposition in NMR and EPR. The power and possibilities of utilizing this quantum property in these fields are numerous. You'll witness how the deliberate manipulation of superpositions allows us to obtain incredibly detailed information about molecular structure, dynamics, and interactions. We'll explore how these powerful tools aid us in medical diagnostics and other fields, all founded on the seemingly paradoxical, yet undeniably real, quantum superposition.

2 Nuclear Magnetic Resonance (NMR)

The nucleus of an atom, often overlooked in our everyday understanding of matter, possesses a fundamental property called spin. This intrinsic angular momentum, quantized in units of /2, behaves like a tiny bar magnet, creating a magnetic moment. When exposed to an external magnetic field, these nuclear spins align either with or against the field, generating a net magnetization. This alignment, however, is not perfectly ordered; thermal fluctuations cause some spins to remain randomly oriented. This is a critical starting point for understanding NMR.

Applying a radio-frequency (RF) pulse, whose frequency precisely matches the energy difference between the spin states (the Larmor frequency), causes a resonant absorption of energy. Nuclei transition from their lower-energy state to the higher-energy state, a phenomenon known as nuclear magnetic resonance. This absorption is not a simple on/off switch, rather it's a complex process influenced by numerous factors, which we'll delve into later. This precise resonance, however, forms the foundation of NMR spectroscopy, allowing us to extract detailed structural information about molecules.

The crucial aspect of the NMR technique lies in the fact that the Larmor frequency is directly proportional to both the strength of the external magnetic field and the gyromagnetic ratio of the nucleus, a unique property for each type of nucleus. Different nuclei within a molecule will experience slightly different magnetic fields due to shielding effects from nearby electrons. These minute variations in local magnetic fields lead to slightly different resonant frequencies, enabling us to distinguish between different types of nuclei and their chemical environments. We're essentially mapping the molecule's magnetic landscape.

Following the RF pulse, the excited nuclear spins don't remain in their higher-energy state indefinitely. They return to their equilibrium state through a process called relaxation. This relaxation occurs via two primary mechanisms: spin-lattice relaxation (T1) and spin-spin relaxation (T2). T1 relaxation describes the return of the bulk magnetization to its equilibrium value, while T2 relaxation reflects the loss of coherence in the spin system. Both relaxation times provide valuable insights into molecular dynamics and the interactions between nuclei. Understanding these relaxation processes is essential for interpreting NMR spectra effectively.

Measuring the response of the nuclei to the applied RF pulses allows us to obtain an NMR spectrum. The spectrum displays a series of peaks or signals, each corresponding to a specific type of nucleus in a distinct chemical environment. The position of a peak on the spectrum (its chemical shift) indicates the shielding experienced by that nucleus. The intensity of a peak reflects the number of nuclei in that environment. Moreover, the peak's shape and width provide information on relaxation times, coupling to neighboring nuclei and molecular motion. The beauty of NMR lies in its versatility. It's not just limited to simple molecules; it can probe the structure and dynamics of complex biological macromolecules such as proteins and nucleic acids. This ability stems from the exquisite sensitivity of NMR to subtle changes in the magnetic environment surrounding nuclei. This means we can observe subtle conformational changes in proteins caused by binding events or even track the dynamic movement of individual atoms within a molecule. The spectrum is a rich tapestry of information, each thread telling a different aspect of the molecule's story.

Sophisticated pulse sequences, beyond the simple single pulse experiment, can be designed to enhance spectral resolution or obtain different types of information. For instance, 2D NMR allows the correlation between nuclei to be revealed, making it an indispensable tool for the complete structure elucidation of organic molecules. Such advanced techniques allow the unfolding of complex molecular structures with unprecedented detail, and provide essential data to various scientific fields including drug discovery and materials science.

NMR finds its place in a vast spectrum of applications. From analyzing the composition of petroleum products to determining the structure of novel pharmaceuticals, NMR spectroscopy's impact is profound. This technique contributes significantly to the field of metabolomics, providing detailed metabolic profiles that are crucial in diagnostics and disease research. In medical imaging, magnetic resonance imaging (MRI) is a powerful, non-invasive tool using NMR principles for generating detailed anatomical images of the human body, greatly impacting healthcare and medical diagnosis.

The power of NMR lies not only in its ability to provide detailed structural information but also in its non-destructive nature. Unlike many other analytical techniques, NMR does not require the destruction or alteration of the sample, allowing for multiple analyses to be performed on the same sample. Its versatility and information richness make it a cornerstone of many scientific disciplines, continuously expanding its range of applications and pushing the boundaries of our understanding of the molecular world. This makes it a vital tool for researchers across the globe.

2.1 Nuclear Spin and Magnetism

Many atomic nuclei possess an intrinsic angular momentum, a property called spin. This spin, often represented by the quantum number I, is quantized, meaning it can only take on specific discrete values. For example, ^{1}H (hydrogen-1) has I = 1/2, while ^{13}C (carbon-13) also has I = 1/2. These spinning nuclei create a tiny magnetic moment, , proportional to their spin. This magnetic moment is what makes them detectable by NMR. Understanding this fundamental connection is crucial for interpreting NMR spectra and for the development of advanced NMR techniques. The interaction between the nuclear magnetic moment and an applied magnetic field is the cornerstone of NMR spectroscopy. A nucleus with a non-zero spin acts like a tiny bar magnet. When placed in an external magnetic field, B, it can align itself either parallel or antiparallel to the field. These two orientations possess slightly different energies, leading to quantized energy levels. The energy difference between these levels is directly proportional to the strength of the applied magnetic field. This energy splitting allows us to manipulate and detect nuclear spins using electromagnetic radiation. The precise energy difference is crucial for resonance and spectral analysis. We will explore the details in the next chapter.

The energy difference between these spin states isn't just dependent on the external magnetic field, but it's also influenced by the local electronic environment. The electrons surrounding the nucleus create their own magnetic fields that interact with the nuclear spin, a phenomenon called chemical shielding. This shielding effect subtly modifies the energy difference, creating variations in resonance frequencies for nuclei in different chemical environments. This slight shift in frequency provides crucial chemical information, enabling us to distinguish between identical atoms in different molecules or locations within a molecule. It's this sensitivity that makes NMR such a powerful analytical tool.

The interaction between the magnetic moment and the external magnetic field isn't static. The nuclei are constantly tumbling and interacting with their surroundings, leading to relaxation processes. These processes are responsible for the return of the excited nuclei to their equilibrium distribution after excitation by a radio frequency pulse. Two key relaxation mechanisms are spin-lattice relaxation (T1) and spin-spin relaxation (T2). T1 reflects how quickly the nuclei lose their excess energy to the surrounding lattice (the sample), while T2 describes the loss of phase coherence among the nuclei due to interactions between them. Understanding these relaxation times provides important insights into molecular dynamics and environment.

Consider an ensemble of nuclei in a magnetic field. At thermal equilibrium, slightly more nuclei are aligned with the field than against it.

The application of a radio frequency (RF) pulse with an energy matching the energy difference between spin states causes transitions between these states—a process called resonance. This resonant absorption of energy is what we detect in NMR spectroscopy. The resulting signal, a free induction decay (FID), is then processed by Fourier transformation to produce the familiar NMR spectrum. The signal's intensity and frequency provide information about the number and environment of nuclei present.

The resonance frequency, often expressed in megahertz (MHz), is directly proportional to the magnetic field strength and the gyromagnetic ratio of the nucleus, a fundamental constant specific to each nuclide. This relationship highlights the importance of both the applied field and the intrinsic properties of the nucleus in the resonance process. Furthermore, this frequency sensitivity is exploited in various NMR techniques for obtaining enhanced spectral resolution and information. The precision of this resonance is key to the powerful applications of NMR. It makes it incredibly effective and widely applied.

The development and application of NMR spectroscopy has led to significant breakthroughs in many scientific disciplines. From determining the structures of complex molecules in organic chemistry to imaging the human body in medicine, NMR has proven to be an indispensable analytical tool. Its sensitivity to the subtle differences in chemical environments makes it exceptionally versatile and precise. This detailed understanding of nuclear spin and its interactions is the basis for all NMR applications. The applications span from medical imaging techniques (MRI) to characterization of materials and much more.

In summary, the interplay of nuclear spin, magnetism, and radio frequencies lies at the heart of nuclear magnetic resonance spectroscopy. The intrinsic magnetic moment of atomic nuclei, their response to external magnetic fields, and the relaxation mechanisms that govern their return to equilibrium are all fundamental concepts essential for understanding NMR. This intricate interplay forms the basis for numerous techniques, offering unique insights into molecular structure, dynamics, and composition across various scientific fields. By understanding the interplay between these phenomena, we unlock the power of NMR for diverse research and industrial applications.

2.2 Resonance Condition

The resonance condition, central to both Nuclear Magnetic Resonance (NMR) and Electron Paramagnetic Resonance (EPR), dictates the precise frequency at which a spin system absorbs energy from an electromagnetic field. Understanding this condition is paramount for successfully employing these powerful spectroscopic techniques. It hinges on the interplay between the magnetic moment of the spin (nuclear or electron), the applied external magnetic field, and the energy of the electromagnetic radiation. In essence, resonance occurs when the energy of the photons in the electromagnetic field exactly matches the energy difference between the spin system's allowed energy levels.

This energy difference is directly proportional to the strength of the external magnetic field (B). A stronger field leads to a larger energy separation, requiring higher-frequency radiation to induce transitions. This relationship is described by the Larmor equation: E = B, where E is the energy difference, is the gyromagnetic ratio (a characteristic constant for each nucleus or electron), is the reduced Planck constant, and B is the external magnetic field strength. The gyromagnetic ratio reflects the sensitivity of the spin to the applied field; larger values indicate greater sensitivity and hence stronger signals.

Practically, this means that to observe resonance, we need to carefully adjust the frequency of the electromagnetic radiation to match the Larmor frequency (= B/2) specific to the nuclei or electrons and the magnetic field strength. This frequency, the resonant frequency, represents the precise point where the energy levels align, allowing the absorption of energy and hence the detection of a signal. Deviations from this precise frequency lead to minimal energy absorption, resulting in a weak or absent signal. Therefore, precision in controlling both the magnetic field and the frequency of the electromagnetic radiation is crucial for optimal NMR and EPR experiments.

Consider, for instance, proton NMR. Protons, the nuclei of hydrogen atoms, possess a specific gyromagnetic ratio. In a typical magnetic field of 14.1 Tesla, the resonant frequency for protons calculates to approximately 600 MHz. If we were to use a lower frequency, the radiation wouldn't have enough energy to bridge the gap between the spin energy levels, and the proton wouldn't absorb the energy efficiently. Similarly, if we were to use a higher frequency, the energy mismatch would yield a weak signal. Therefore, precise control and calibration are essential to achieve resonance.

The significance of understanding the resonance condition extends beyond simply obtaining a signal. It allows us to selectively target specific nuclei or electrons within a complex sample. For example, in a molecule with multiple types of nuclei (like ^{1}H, ^{13}C, and ^{1}N), each isotope has a unique gyromagnetic ratio. By adjusting the radio frequency appropriately, we can selectively excite and detect signals only from one type of nucleus at a time, simplifying complex spectra and providing detailed information about molecular structure. This technique, often referred to as selective excitation, is routinely used in advanced NMR techniques. Furthermore, differences in the local magnetic field surrounding each nucleus due to the electron cloud (chemical shielding) cause slight variations in the resonance frequency, providing valuable information about chemical environment and bonding.

In EPR, the principle remains the same. However, the gyromagnetic ratio of the electron is considerably larger than that of most nuclei. As a consequence, electron paramagnetic resonance experiments are often conducted at higher frequencies in the microwave region, typically in the gigahertz range, even for moderate magnetic field strengths. This higher sensitivity to magnetic fields allows for the detection of unpaired electrons in various samples, making EPR a powerful tool for studying paramagnetic species and their interactions with their surroundings.

In conclusion, the resonance condition is the bedrock upon which NMR and EPR spectroscopy are built. A thorough grasp of this principle is crucial for any researcher or student aspiring to use these powerful techniques for studying molecular structure, dynamics, and electronic properties. The ability to precisely control and manipulate the frequency of the electromagnetic radiation and the magnetic field strength underpins the success of NMR and EPR experiments and their applications in diverse scientific disciplines. Its practical implications are far-reaching, impacting the fields of medicine, materials science, and chemical analysis. Therefore, a deep understanding of this phenomenon is fundamental to the successful application of these techniques. The precise manipulation and understanding of the resonance condition lead to a precise understanding of quantum phenomena in magnetic resonance.

2.3 Relaxation Mechanisms

Spin-lattice relaxation, often denoted as T1 relaxation, describes the return of excited nuclear spins to their equilibrium Boltzmann distribution through interactions with the surrounding lattice. Imagine the spins as excited dancers, gradually losing their energy to the relatively still audience (the lattice) until they settle back into their usual, less energetic positions. This energy transfer occurs through various mechanisms, significantly fluctuating local magnetic fields caused by molecular motion, which effectively disrupt the spin's precessional motion around the external magnetic field. The rate of this energy dissipation depends on the correlation time of these molecular motions—the average time between changes in the magnetic field environment—and varies greatly depending on factors like temperature, viscosity, and the molecular structure itself. This dependence is crucial for understanding the dynamics of molecules in solution and solids. Faster molecular motion usually leads to shorter T1 relaxation times, a fundamental principle utilized extensively in structural and dynamic analyses of molecules.

T2 relaxation, or spin-spin relaxation, is a different process that doesn't involve energy exchange with the lattice. Instead, it's a loss of coherence amongst the spins themselves. Picture a group of synchronized swimmers; initially, they move in perfect unison, but over time, slight variations in their movements lead to a loss of the overall synchronized pattern. In NMR, this "loss of synchronization" represents the decay of the transverse magnetization—the net magnetization component perpendicular to the external magnetic field. The major cause of this dephasing is the presence of slight variations in the local magnetic fields experienced by each individual spin. These variations, stemming from the inherent randomness of molecular motion and the interaction between spins, cause the spins to precess at slightly different rates, leading to a gradual loss of phase coherence and the decay of the transverse magnetization. The rate of this dephasing dictates the T2 relaxation time, influencing the linewidth of spectral lines—broader lines indicate faster T2 relaxation.

The relationship between T1 and T2 relaxation times provides valuable insights into molecular dynamics. Typically, T1 is longer than T2. This is because T1 is only influenced by the fluctuations that directly influence the energy of the spin system. T2 however, is affected by these fluctuations, and additionally, by any inhomogeneity in the magnetic field that cause the spins to dephase even without energy transfer. The difference between T1 and T2 relaxation times can be significant and offers valuable information about the degree of magnetic field homogeneity and the efficiency of spin-spin interactions, thus impacting NMR spectral resolution and signal strength. A meticulous analysis of these relaxation times allows researchers to decipher detailed information about molecular structures, dynamics, and their environments.

Dipolar interactions, where magnetic dipole moments interact with each other, exert a substantial influence on relaxation. Imagine two tiny magnets positioned close to one another. They'll exert forces on each other, affecting their orientation and leading to energy exchange or dephasing. In NMR, this occurs between nuclear spins, causing changes in their energy levels and affecting their precessional motion, thus influencing both T1 and T2 relaxation times. The strength of this interaction depends inversely on the cube of the distance between the spins; therefore, it's most effective at short distances. Understanding the dynamics of these dipolar interactions is essential in determining molecular conformation and internuclear distances. Chemical exchange, or the rapid swapping of atoms between different molecular environments, also impacts relaxation times. The rates of chemical exchange significantly influence spectral lineshapes and relaxation processes, providing insights into reaction mechanisms and dynamic equilibria in chemical systems. Quadrupolar relaxation is specific to nuclei with a spin quantum number greater than 1/2, which possess a non-spherical charge distribution. Unlike dipolar relaxation, quadrupolar relaxation originates from the interaction of this non-spherical nuclear charge distribution with the fluctuating electric field gradients caused by the surrounding molecules. These fluctuations induce transitions between the nuclear spin energy levels and hence contribute to T1 relaxation. The rate of this relaxation is highly sensitive to the symmetry and dynamics of the molecular environment surrounding the nucleus. The presence of quadrupolar nuclei in a molecule considerably shortens the T1 and T2 relaxation times, often leading to broader lines in NMR spectra. Careful analysis of these effects helps in understanding the molecular structure and dynamics, particularly for molecules containing quadrupolar nuclei like deuterium, nitrogen-14 and several other important elements. Understanding these mechanisms is crucial for successfully interpreting NMR spectra. The study of relaxation mechanisms is not just an academic exercise but a cornerstone in the advancement of many NMR applications.

3 Electron Paramagnetic Resonance (EPR)

Electrons, unlike their nuclear counterparts, possess an intrinsic angular momentum, or spin, which is also a quantized property. This spin, coupled with their charge, makes them exquisitely sensitive to magnetic fields, a foundation for Electron Paramagnetic Resonance (EPR) spectroscopy. Unlike NMR, which probes the nuclei of atoms, EPR focuses specifically on unpaired electrons, those with a net magnetic moment. This means EPR is uniquely suited for studying systems with paramagnetic centers, such as transition metal ions, free radicals, and defects in materials. These unpaired electrons interact with the magnetic field in a predictable manner, enabling detailed investigations into their environments.

The basic principle underlying EPR is similar to NMR: the application of a magnetic field splits the electron's spin energy levels, and irradiation with electromagnetic radiation of a specific frequency—the resonant frequency—can induce transitions between these levels. This absorption of energy is then detected, providing a spectrum that reflects the electron's environment and interactions. The sensitivity of EPR makes it a powerful tool for investigating a wide range of phenomena.

A crucial aspect differentiating EPR from NMR lies in the hyperfine interaction. This interaction arises from the magnetic moment of the electron interacting with the magnetic moments of nearby nuclei. These nuclear spins influence the electron's resonant frequency, producing a characteristic splitting pattern in the EPR spectrum. The magnitude of this splitting offers valuable information about the number and identity of surrounding nuclei, providing a detailed picture of the electron's immediate environment. This is a powerful tool for understanding molecular structure and dynamics, particularly in free radical systems.

The experimental setup for EPR involves a strong magnetic field, a microwave source for irradiation, and a sensitive detector to measure the absorbed energy. The sample is placed within the magnetic field, and the microwave frequency is swept, recording the absorption signal as a function of the magnetic field strength. The resultant spectrum shows characteristic peaks corresponding to the resonant frequencies of the unpaired electrons. The position and shape of these peaks are influenced by various factors, including the applied magnetic field, the nature of the unpaired electron, and its interactions with the surrounding environment.

Applications of EPR extend across numerous scientific disciplines. In chemistry, EPR is employed to study reaction mechanisms, identify short-lived intermediates, and characterize free radicals involved in biological and chemical processes. For example, researchers use EPR to study oxidation processes in living systems or the formation of reactive species in chemical reactions. Materials science benefits greatly from EPR as a means to characterize defects in materials, providing insights into their structure and influencing their properties. Studying semiconductors or catalysts provides an excellent example of materials science applications of EPR.

Beyond these applications, EPR plays a role in medicine, though not as widely as NMR. EPR spectroscopy, particularly spin trapping, allows researchers to identify and quantify free radicals in biological systems, crucial in studying oxidative stress and disease mechanisms. This field holds great promise for future advancements in understanding the role of free radicals in health and disease. While often overshadowed by NMR in terms of widespread applications, EPR's unique ability to probe unpaired electrons gives it a crucial place in the arsenal of spectroscopic tools for the investigation of matter at the molecular level.

The information provided by EPR spectra is rich and nuanced, requiring sophisticated analysis techniques to extract meaningful data. Simulations, often based on advanced theoretical models, help in interpreting the spectra and elucidating the molecular structures and dynamics. Combining EPR with other spectroscopic techniques, such as NMR or UV-Vis, often provides complementary information that enhances the understanding of complex systems. Understanding the principles behind EPR and its experimental techniques provides a solid foundation for advanced research endeavors.

This comprehensive understanding of EPR's strengths, coupled with careful experimental design and data analysis, unlocks its considerable potential for unlocking new knowledge and advancing scientific inquiry across various scientific disciplines. The development of new techniques and instrumentation constantly enhances the capabilities of EPR, pushing the boundaries of our understanding of the molecular world, and contributing to advancements in materials science, chemistry, and biology. Mastering EPR is a significant step for any aspiring scientist involved in the study of molecules and materials.

3.1 Electron Spin and Magnetism

Electrons, unlike their nuclear counterparts, possess an intrinsic angular momentum called spin, a fundamental quantum property that's crucial for understanding magnetism. This inherent spin, often visualized as a tiny rotating charge, generates a magnetic moment, a vector quantity pointing in a specific direction. This magnetic moment interacts strongly with external magnetic fields, leading to observable phenomena exploited in Electron Paramagnetic Resonance (EPR) spectroscopy. Unlike the relatively weak magnetic moments of nuclei, electron spins exhibit a much stronger interaction with magnetic fields due to their significantly smaller mass. This difference makes EPR spectroscopy a powerful tool for studying systems with unpaired electrons, including free radicals, transition metal ions, and defects in materials.

The interaction between an electron's magnetic moment and an external magnetic field is quantized. This means the electron spin can only adopt specific orientations relative to the field, usually described as "spin up" and "spin down," corresponding to different energy levels. Applying a resonant microwave frequency, precisely tuned to match the energy difference between these levels, induces transitions between these spin states. The absorption of this microwave energy forms the basis of the EPR spectrum, a unique fingerprint of the electron's environment. Analyzing these spectra provides crucial information about the molecular structure, dynamics, and electronic properties of the system under investigation.

Furthermore, the electron spin's interaction with its surroundings is complex. The magnetic field experienced by the electron isn't just the externally applied field; it's also influenced by the nuclear spins within the molecule. This is known as hyperfine interaction. These interactions, though subtle, split the electron spin energy levels, creating intricate patterns in the EPR spectrum that reveal details about the nuclear environment surrounding the unpaired electron. The strength and number of these hyperfine splittings depend on the number and types of nuclei interacting with the electron, their distance, and their nuclear spins. Mastering the interpretation of these hyperfine patterns is essential for unraveling molecular structures and understanding the dynamics of electron spin interactions.

Unlike NMR which focuses primarily on diamagnetic substances, EPR's sensitivity to unpaired electrons makes it ideal for studying paramagnetic species. This allows EPR to address unique research questions, complementing NMR's capabilities and providing insights inaccessible through other spectroscopic methods. For instance, EPR finds applications in studying reactive intermediates in chemical reactions, understanding the structure and function of metalloproteins, characterizing defects in semiconductors, and investigating the behavior of radicals in biological systems. The diversity of applications underscores the power and versatility of EPR in exploring a wide range of scientific and technological fields.

The analysis of EPR spectra, however, requires a thorough understanding of quantum mechanics and spin physics. While the basic principles involve relatively straightforward concepts, the detailed interpretation of complex spectra requires expertise and advanced analytical techniques. This necessitates a strong foundation in quantum mechanics, familiarity with the mathematical formalism of spin operators, and proficiency in using specialized EPR software for spectral simulation and analysis. This chapter serves as an introduction to these concepts and aims to provide you with sufficient background for interpreting the basic features of EPR spectra and appreciating the sophisticated information they can yield. Remember, practice is key. Work through examples, consult relevant literature, and don't hesitate to explore advanced resources once you have a firm grasp of the fundamentals. By understanding electron spin and its magnetic interactions, you'll be well-equipped to explore the fascinating world of EPR spectroscopy and its many applications. The more you practice the clearer the picture will become.

The strength of the magnetic field experienced by the electron is crucial in EPR. This effective field is not simply the externally applied field; it's modified by the local electronic environment. Factors such as the electron's orbital angular momentum, the presence of other paramagnetic centers, and crystal field effects can all influence the effective magnetic field and therefore the resonant frequency needed to induce spin transitions. This necessitates precise control over the applied magnetic field and careful consideration of the sample environment. Furthermore, understanding these environmental factors and their impact on the observed EPR spectrum is crucial for accurate interpretation of the data and extracting meaningful insights into the studied material. For example, subtle variations in the crystal structure surrounding a paramagnetic ion can lead to observable changes in the EPR spectrum. Understanding the detailed interactions between the electron spin and its environment is crucial to interpreting EPR spectra accurately. These interactions significantly influence the energy levels of the electron spin and, consequently, the resonant frequencies at which spin transitions occur. In essence, the spectrum is not solely a reflection of the electron's spin; it provides a comprehensive "fingerprint" of its surrounding environment. Mastering the subtleties of spin-environment interactions will significantly enhance your ability to interpret complex EPR data and extract valuable information about the system being studied. Moreover, exploring these details will expand your understanding of magnetism at the atomic and molecular levels. The precise analysis of hyperfine structure, for instance, allows for the determination of nuclear spin and thus identification of the specific isotopes in the molecule under study. The practical applications of EPR spectroscopy are incredibly diverse, ranging from basic research to industrial analysis. In material science, EPR is used to characterize defects and impurities in semiconductors, a crucial aspect of optimizing material properties and performance. In chemistry, EPR aids in studying reaction mechanisms by detecting short-lived radicals and intermediates. Moreover, EPR finds applications in biochemistry, enabling the study of metalloenzymes, understanding their active sites, and observing the dynamics of electron transfer processes. The versatility of EPR as an analytical technique makes it invaluable across various scientific disciplines. Its ability to probe the electronic structure of materials and molecular systems non-destructively allows for both qualitative and quantitative information.

3.2 Hyperfine Interactions

The hyperfine interaction, a subtle yet powerful phenomenon, arises from the magnetic interaction between a nucleus and an unpaired electron. Unlike the relatively strong interactions governing electron spin resonance (ESR) itself, the hyperfine interaction is a much smaller effect, often treated as a perturbation. This interaction provides incredibly valuable structural information, enabling us to extract detailed knowledge about the environment surrounding the unpaired electron. For example, by analyzing the hyperfine splitting pattern, we can determine the number of interacting nuclei and their distance from the electron. This information is invaluable in various fields such as materials science, chemistry and biology, where determining molecular structure is paramount.

Imagine the electron, a tiny spinning magnet, residing within a molecule. Nearby, a nucleus, another small magnet, exerts its own influence. This mutual magnetic interaction leads to a splitting of the electron spin energy levels. The magnitude of this splitting is directly proportional to the strength of the hyperfine interaction, and therefore to the nuclear magnetic moment and the distance between the electron and nucleus. Different isotopes of the same element will have different magnetic moments and hence different hyperfine splittings, providing a powerful tool for isotopic identification within a molecule.

Experimental observation of hyperfine interactions usually involves careful analysis of the EPR spectra. In simple cases, a single nucleus will interact with the unpaired electron, resulting in a characteristic splitting of the ESR line into a doublet. More complex situations arise when multiple nuclei interact, leading to more intricate splitting patterns, often described using the rules of combinatorics. For instance, two equivalent nuclei lead to a triplet, three equivalent nuclei result in a quartet and so on, with the intensity of the lines following Pascal's triangle. Analyzing these patterns allows us to determine the number and type of nuclei interacting with the unpaired electron, even within complex molecules.

These complex patterns are not always easily deciphered. Isotropic hyperfine interactions are relatively straightforward to understand. But anisotropic interactions, which depend on the orientation of the molecule in the magnetic field, add a layer of complexity to the spectral analysis. Detailed simulations and fitting of the experimental spectra are often required, sometimes employing sophisticated software tools. This analysis can reveal a wealth of information about the electronic structure and molecular geometry.

Consider a situation where the unpaired electron is part of a radical molecule. The hyperfine coupling constants, which directly reflect the strength of the hyperfine interaction, provide crucial insights into the electronic density distribution within the molecule. Large coupling constants signify a high degree of electron density near a particular nucleus, implying a strong interaction. Conversely, small coupling constants suggest a lower electron density and a weaker interaction. Therefore, the entire picture of the electron density is revealed through careful scrutiny of these constants, which can be used to understand the molecule's reactivity and dynamics.

This information is not merely theoretical. Hyperfine interactions find applications in various fields. In material science, the technique is invaluable for characterizing defects in crystals and semiconductors. By probing the unpaired electrons associated with these defects, we can accurately determine their nature and their impact on the material properties. In chemistry, hyperfine interactions help establish the structure of organic radicals and transition metal complexes. These details are critical in understanding their reactivity, which drives catalytic processes or impacts biological processes.

Furthermore, hyperfine interactions are of great significance in biology, where they aid in understanding the structure and function of metalloproteins. These proteins, containing metal ions, play critical roles in various biological processes. The hyperfine interactions involving these metal centers offer essential insights into the electronic structure and reactivity of these vital biomolecules, leading to a better comprehension of enzyme mechanisms and metabolic pathways.

The analysis of hyperfine interactions can be challenging but rewarding. It requires a strong understanding of quantum mechanics and spectral analysis. However, with the right tools and techniques, it provides access to a wealth of information about molecular structure and dynamics that is unavailable through other spectroscopic methods. The power of this technique lies in its ability to reveal detailed information about the environment of unpaired electrons, allowing us to probe the intricacies of complex systems. Mastering the interpretation of hyperfine interactions is therefore a crucial skill for any aspiring researcher in the fields of chemistry, physics, and materials science, opening the door to a universe of intricate molecular details. Further exploration into advanced techniques and computational methods provides more tools to resolve even the most complex hyperfine structures, paving the way for continued innovation and understanding.

4 Quantum Entanglement and NMR

Nuclear magnetic resonance (NMR) provides a powerful platform for exploring the bizarre world of quantum entanglement. We can create and manipulate entangled states, the very essence of quantum weirdness, within the controlled environment of an NMR spectrometer. This allows us to directly observe and even utilize the strange correlations predicted by quantum mechanics, defying classical intuition. Imagine two particles, linked in such a way that knowing the state of one instantly tells you the state of the other, regardless of the distance separating them. This isn't science fiction; it's quantum entanglement, and NMR lets us study it firsthand.

The simplest entangled states we can create are Bell states, named after John Bell, a physicist who played a crucial role in solidifying the reality of entanglement. These states, involving two qubits (quantum bits), are fundamentally different from any classical system. In a classical system, you could think of each qubit existing independently, in a defined state, but a Bell state forces a unique relationship. Measuring one qubit's state instantly determines the other's, a phenomenon deeply connected to the superposition principle explored earlier. NMR's precise control over nuclear spins makes the generation and manipulation of these Bell states remarkably straightforward.

The ability to generate and manipulate entangled states within an NMR system opens up exciting possibilities in quantum computing. NMR quantum computers, while not as scalable as some other approaches, provide a valuable proving ground for quantum algorithms. Researchers have demonstrated basic quantum algorithms like the Deutsch-Jozsa algorithm, highlighting the computational power of entangled states in solving specific types of problems more efficiently than classical computers. This experimental platform allows a more concrete understanding of the principles of quantum computation, paving the way for future, more powerful quantum computers.

Beyond quantum computing, the entangled states generated through NMR find applications in a diverse array of fields. Quantum teleportation, a concept often portrayed in science fiction, utilizes entangled pairs to transfer quantum information from one location to another without physically transporting the qubit. NMR experiments have demonstrated rudimentary forms of quantum teleportation, showcasing the potential of this technology for secure communication and quantum networking. Bell inequalities, mathematical expressions that distinguish quantum correlations from classical ones, can be tested using NMR experiments. Violations of these inequalities provide compelling experimental evidence supporting the reality of entanglement, surpassing classical explanations. These tests serve as a critical component in upholding the validity of quantum mechanics, constantly challenging our understanding of reality at its most fundamental level. They are not just theoretical constructs but experimentally verifiable aspects of the quantum world, explored through the precise tools of NMR. NMR's ability to perform such experiments accurately has greatly helped in advancing the field of quantum physics.

NMR's role in studying entanglement isn't merely limited to fundamental research. The unique correlations arising from entangled states have the potential to enhance sensitivity in various applications, such as quantum sensing. Imagine sensors that can detect incredibly small changes in magnetic fields or other physical quantities, far beyond the capabilities of classical sensors. Entanglement-enhanced NMR techniques are already being explored to achieve these goals, opening doors for advancements in medical imaging, materials science, and beyond. This advancement may greatly impact a wide range of industries.

The study of quantum entanglement using NMR techniques is a rapidly evolving field, continually revealing new applications and deepening our understanding of the quantum world. As our ability to control and manipulate entangled states improves, so too will our ability to harness the power of quantum mechanics for scientific advancements and technological innovation. The precision and control offered by NMR spectrometers make them invaluable tools in this endeavor, promising a future where the unique properties of entanglement are utilized to solve complex problems and develop cutting-edge technologies. The potential impact on many industries and scientific research is enormous. The continued exploration of quantum phenomena within the framework of NMR promises a fruitful and fascinating journey into the future.

4.1 Bell States and NMR

Bell states, the quintessential examples of entangled quantum states, find a particularly elegant application within the realm of nuclear magnetic resonance (NMR). These maximally entangled two-qubit states, denoted as |⟩, |⟩, |⟩, and |⟩, represent a profound departure from classical physics; their properties defy intuitive understanding and offer unique opportunities for quantum information processing. In NMR, we leverage the inherent quantum nature of nuclear spins to create and manipulate these Bell states, opening doors to experiments demonstrating the foundational principles of quantum mechanics and exploring the potential of quantum computation.

The creation of Bell states in NMR typically involves a series of carefully calibrated radio-frequency pulses applied to coupled nuclear spins. Consider a system with two coupled spins, such as those found in a molecule like 1H-13C. Starting from a well-defined initial state, a sequence of pulses rotates these spins, interweaving their quantum properties to create the desired entanglement. This is a complex process, demanding precise control over pulse duration, amplitude, and phase. Imperfect pulse shaping or unwanted interactions with the surrounding environment lead to decoherence—the loss of entanglement—which is one of the significant challenges in implementing quantum algorithms.

The experimental verification of Bell states depends on analyzing the spin states of the involved nuclei after the entanglement process. Measurement techniques such as spectroscopy detect the populations and coherences of the spin states, allowing us to infer whether entanglement has been successfully generated and maintained. The specific techniques employed are tailored to the particular system under consideration and may involve a combination of pulsed NMR methods, careful selection of nuclear isotopes, and data processing techniques that suppress noise and refine the signals.

One of the most compelling aspects of using NMR to generate and manipulate Bell states is its relative accessibility compared to other quantum computing platforms. The technique benefits from well-established experimental protocols, allowing researchers to focus more readily on quantum algorithms and less on the intricacies of device fabrication. However, the scalability of NMR quantum computation faces limitations because of the limited number of qubits attainable in NMR systems. This constraint arises from the reliance on the macroscopic ensemble average of numerous identical molecules, rather than individual quantum systems, which often limits the number of qubits that can be effectively addressed and controlled.

Despite these limitations, the accessibility of NMR systems makes them invaluable tools for educational purposes. Undergraduate and graduate students can directly engage with the fundamental principles of quantum entanglement through hands-on experiments. By directly manipulating Bell states, they gain a far deeper and more intuitive understanding of quantum superposition, interference, and the intricacies of quantum measurement. Such experiments are essential in bridging the gap between the theoretical concepts and the practical reality of quantum phenomena.

Beyond their role in demonstrating entanglement, Bell states also serve as building blocks for more complex quantum algorithms implemented in NMR. For instance, the creation of Bell states forms the basis of quantum teleportation protocols, where quantum information is transferred between two spatially separated qubits, a phenomenon that showcases the non-classical correlations at the heart of entanglement. While challenges still exist in scaling NMR-based quantum computing for practical applications, the ability to generate and manipulate Bell states within the well-established framework of NMR spectroscopy provides an accessible platform for exploring the foundational principles and applications of quantum information science.

Understanding the methods for creating and detecting Bell states in NMR involves delving into the detailed theoretical frameworks of spin dynamics and quantum information theory. While this may initially seem daunting, focusing on the fundamental principles and building upon a solid grasp of basic NMR techniques is key to understanding the core concepts of this field. As you learn, you'll begin to appreciate the power of NMR as a uniquely accessible and practical technique to explore the fascinating world of quantum entanglement. This approach provides a direct pathway for students to engage with these otherwise abstract ideas, translating complex theoretical concepts into tangible experimental results.

Many research teams continue to actively explore the possibilities of NMR-based quantum computing, striving to improve scalability and coherence times. Current efforts focus on developing new pulse sequences for higher-fidelity quantum gates, creating novel approaches to controlling noise and decoherence, and finding innovative ways to scale the number of qubits. The integration of advanced techniques from other areas of quantum information science, like quantum error correction, promises to further enhance the potential of NMR in the field of quantum computing. Furthermore, researchers are exploring the use of NMR to probe complex molecular systems, exploiting entanglement for improved resolution and sensitivity in the analysis of molecular structures. This opens the door for novel applications in materials science, chemistry, and drug discovery.

4.2 Quantum Computing with NMR

Nuclear Magnetic Resonance (NMR) offers a unique and accessible platform for exploring the principles of quantum computing. Unlike other approaches requiring extreme cryogenic temperatures and highly specialized equipment, NMR experiments can be conducted at room temperature using readily available laboratory instruments. This accessibility makes it an ideal pedagogical tool for introducing fundamental concepts of quantum computation to students. The basic premise lies in leveraging the quantum properties of nuclear spins within molecules to perform computations. We can manipulate these spins using precisely controlled radio-frequency pulses, creating and manipulating quantum states.

These manipulations, analogous to logic gates in classical computing, allow us to perform computations on the quantum level. A crucial aspect is the ability to create coherent superpositions of nuclear spin states, a key feature of quantum algorithms. This means we aren't just dealing with spins being up or down; we deal with spins existing in a probabilistic mix of both simultaneously. We can then precisely engineer these probabilistic mixtures with pulses. It's this ability to be in multiple states at once that allows quantum computers to potentially outperform classical computers for specific tasks.

The process involves preparing a sample, typically a liquid containing a molecule with suitable nuclei (e.g., ^{1}H, ^{13}C, or ^{1}N), and placing it within a strong magnetic field. Then, carefully timed radio-frequency pulses applied through the coil surrounding the sample selectively excite the spins and initiate the quantum computation. The quantum computation progresses through a series of such precisely controlled pulses and the evolution under the influence of the magnetic field. Finally, we measure the resulting spin states.

However, NMR quantum computing faces challenges. A major limitation stems from the fact that we're not working with a single quantum system, but rather an ensemble of many identical molecules. The measured signal is an average over the behavior of all the molecules, resulting in a signal-to-noise ratio significantly lower than what is achievable in other quantum computing approaches. This ensemble averaging dilutes the quantum behavior, which can be a real hurdle. This means the signals are much weaker, making it more challenging to extract the computational results accurately.

Another crucial aspect is the scalability. While basic quantum algorithms have been successfully implemented on relatively small numbers of qubits (using nuclear spins as qubits), scaling up to larger numbers of qubits in NMR poses significant technical hurdles. The complexity of pulse sequence design and the challenge of maintaining coherence increase exponentially with the number of qubits. This is a significant bottleneck for building larger and more powerful quantum computers.

Despite these limitations, NMR quantum computing remains a valuable tool. It serves as a practical platform to experimentally demonstrate fundamental quantum computing algorithms and to understand the behavior of entangled quantum states. The relatively straightforward experimental setup makes NMR an exceptionally strong tool for educational purposes, allowing students to gain hands-on experience in implementing and studying quantum algorithms, providing a valuable introduction to this complex area. Its strength lies in its affordability and conceptual clarity compared to other quantum computing techniques, especially for introductory learning. Researchers also continue investigating techniques to overcome the scaling challenges and to increase the signal-to-noise ratio, constantly improving its potential for more complex computations.

The development of more sophisticated pulse sequences, innovative experimental methods, and improvements in signal processing techniques continue to push the boundaries of what is possible with NMR quantum computing. While it might not replace other quantum computing approaches entirely, its unique advantages make it a valuable and powerful tool in the quantum computing toolbox, particularly for education and exploring fundamental quantum phenomena in a practical and easily accessible setting. While more advanced techniques offer higher fidelity, NMR's accessibility makes it an excellent starting point for understanding the basic principles underlying this rapidly developing field. It allows for a solid foundation before moving to more intricate systems.

4.3 Entanglement Applications

Harnessing the unique properties of entangled states offers exciting possibilities in NMR. Imagine two nuclear spins, perfectly intertwined - a change in one instantly affects the other, no matter the distance separating them. This bizarre connection, fundamentally quantum, allows for innovative approaches to computation and sensing previously unachievable.

Entanglement's power lies in its ability to create highly correlated systems. These correlations surpass anything classical physics can predict. We can exploit this correlation to enhance the signal-to-noise ratio in NMR experiments. By entangling multiple nuclear spins, we can collectively measure their properties, effectively amplifying the signal and allowing for the detection of weaker signals, crucial in studying low-concentration analytes or observing subtle molecular dynamics. This improvement bypasses limitations imposed by traditional NMR methods, enabling more precise measurements and detailed analyses of complex systems. The ability to perform experiments with improved sensitivity has huge applications across many areas of science, such as in studying chemical reactions or biological processes in real-time. This offers new opportunities to design better sensors and improve experimental methods. Imagine developing ultra-sensitive detectors for trace amounts of toxins or medical biomarkers, opening doors for earlier and more accurate diagnoses.

Consider quantum computing, a field heavily reliant on entanglement. NMR quantum computers, though not as scalable as some other approaches, provide a valuable platform for exploring fundamental quantum algorithms and demonstrating the potential of entanglement-based computations. By utilizing entangled nuclear spins as qubits, the basic units of quantum information, we can perform calculations that are impossible for even the most powerful classical computers. Although scaling up NMR quantum computers faces challenges, their value in quantum algorithm development and educational purposes remains substantial. The ability to manipulate entangled states in NMR systems is a significant step towards building a more powerful quantum computer, with entanglement as the key ingredient. Researchers are now developing methods to create larger and more complex entangled states within these systems.

One intriguing application lies in quantum metrology. Entanglement allows the creation of highly sensitive sensors for physical quantities like magnetic fields or temperature. By entangling multiple nuclear spins, we dramatically improve the precision of measurement beyond the standard quantum limit. The correlated nature of entangled systems allows for an improved precision in the experiments that directly measure the amount of a certain physical quantity. This is relevant in diverse areas like medical imaging, where more accurate measurements lead to improved diagnostics, or material science, where identifying subtle material properties unlocks the development of advanced materials with novel applications. Such advancements could revolutionize fields relying on precise measurements, including navigation systems, and fundamental physics research.

Entanglement is crucial in simulating complex quantum systems. Creating entangled states mirroring the interactions within a molecule allows for simulating its properties with far greater accuracy than classical computational methods can achieve. This approach unlocks detailed insights into molecular dynamics, reaction mechanisms, and other complex processes, leading to breakthroughs in drug design, materials discovery, and our overall understanding of fundamental processes in nature. Simulating quantum systems in NMR has improved our understanding of molecular properties and reaction dynamics.

Finally, while challenges remain in scaling up entanglement-based NMR applications, ongoing research continues to refine control techniques and expand the capabilities of entanglement in NMR. Improved control over entangled states paves the way for more sophisticated applications, potentially revolutionizing diverse fields. New techniques and experimental methodologies continuously improve our ability to manipulate and utilize entangled nuclear spins. Further investigation into the dynamics of entangled states in more complex systems is crucial. This pushes the boundaries of quantum physics and offers numerous possibilities for future development.

4.4 Quantum Teleportation

Quantum teleportation, a concept seemingly ripped from science fiction, is a remarkably real application of quantum entanglement. It doesn't involve the transportation of matter; instead, it's the transfer of a quantum state from one particle to another, instantaneously, even across vast distances. Think of it as photocopying a quantum state, not the physical particle itself. This is achieved by leveraging the intrinsically linked fates of entangled particles.

The process begins with two entangled particles, often photons, A and

B. Particle A is carefully prepared in a known entangled state with particle B. Importantly, a third particle, C, carries the quantum state we wish to teleport; this is the particle we want to "copy." Crucially, we must perform a joint measurement on particles A and C. This measurement, inherently probabilistic, doesn't reveal the specific state of C, but it entangles A and C in a new, correlated way. This combined measurement collapses the wave function of the entangled pair, influencing the state of B remotely.

The nature of this entanglement is such that once A and C are measured, the original quantum state encoded on particle C is now inexplicably found on particle B. It's a remarkable feat, seemingly violating classical principles of information transfer. The information about the state hasn't traveled faster than light; rather, the correlation between A and B acted as an instant conduit, transferring the quantum information. The original state on particle C is destroyed in the process, adhering to the no-cloning theorem of quantum mechanics. It is not duplicated, it's transferred.

Importantly, the teleportation process requires prior entanglement. You can't just conjure up entangled particles at will. Creating entangled states involves sophisticated methods and often requires specialized equipment and significant effort. The entangled pair acts as a channel for the quantum information, and thus creating this channel is the pre-requisite of the process. The success rate of quantum teleportation isn't perfect; probabilities and quantum noise significantly impact experimental outcomes. This limitation highlights the ongoing challenges of achieving high fidelity teleportation in real-world scenarios.

This limitation isn't just a matter of technological refinement; it highlights the inherent probabilistic nature of quantum mechanics. The measurement on A and C influences the state of B probabilistically. The successful outcome must then be verified, further increasing the complexity. While the concept is theoretically elegant, the practical implementation is far from trivial. We're talking about incredibly delicate quantum systems, highly susceptible to environmental interference. The slightest disruption can completely derail the teleportation process, leading to incorrect state transfers.

Despite the challenges, quantum teleportation is not a mere theoretical curiosity. It holds immense potential for quantum computing. By using this technique, it becomes possible to move qubits, the fundamental units of quantum information, between different parts of a quantum computer. This allows for the construction of more robust and complex quantum circuits, a crucial step toward building powerful quantum computers. It facilitates a kind of modularity within a quantum computer that isn't possible with classical systems. The ability to "move" information without physically moving qubits reduces errors and simplifies the design and construction of quantum processors.

Furthermore, advancements in quantum teleportation have significant implications for quantum communication networks. Imagine a secure quantum internet, where information is transmitted in a way that is fundamentally resistant to eavesdropping. Quantum teleportation provides a pathway toward this, albeit one that requires significant further development. It remains one of the building blocks that is crucial to a future quantum communications infrastructure. Developing better entanglement sources and error correction codes are crucial steps toward realizing a practical quantum internet.

The future of quantum teleportation is brimming with possibility. Ongoing research focuses on improving the fidelity and efficiency of the process, increasing its robustness against noise and environmental effects. Developing more stable entanglement sources is paramount. Also, exploring new teleportation protocols to increase the efficiency and reduce resource requirements will play a vital role. The integration of quantum teleportation within larger quantum networks is a major challenge, requiring efficient methods for distributing and managing entanglement across multiple nodes. The ultimate goal is to create a large-scale, fault-tolerant quantum network, with teleportation as a core enabling technology. This would revolutionize computation, communication, and various other fields. It's not just about transferring quantum states; it's about creating a fundamentally new way of interacting with the quantum world, opening doors to previously unimagined technologies.

4.5 Bell Inequalities

Bell inequalities, stemming from the work of John Bell, provide a crucial test for the validity of local realism against quantum mechanics. Local realism assumes that physical systems possess definite properties independent of measurement and that influences cannot travel faster than light. Quantum mechanics, however, predicts correlations between entangled particles that violate these assumptions. These violations, experimentally confirmed, showcase the profound differences between the classical and quantum worlds. Understanding Bell inequalities is crucial for grasping the non-classical nature of quantum entanglement.

Let's consider a simple scenario involving two entangled photons. Their polarizations, for example, are correlated. If one is measured to be vertically polarized, the other will inevitably be horizontally polarized, and vice-versa. Classical physics, adhering to local realism, can account for correlations to a certain extent, but there are limits. Bell inequalities set mathematical boundaries on the strength of correlations permissible under local realism.

These inequalities are expressed as mathematical expressions. These formulas define relationships between probabilities of different measurement outcomes on the entangled particles, limiting the possible correlations. If experimental results violate these inequalities, it directly implies that the assumptions underpinning local realism - locality and realism - are incorrect. Crucially, many experiments have, in fact, shown violations of Bell inequalities, strongly suggesting that the quantum world operates in a way fundamentally different from our classical intuitions.

The experimental verification of Bell inequality violations has far-reaching implications. It decisively refutes the possibility of a local realistic description of quantum mechanics. This has significant implications for our understanding of fundamental physics, particularly the interpretation of quantum mechanics. The implications extend beyond theoretical physics and directly impact the development of quantum technologies.

Consider the case of quantum cryptography. The secure communication protocols rely on the non-local correlations of entangled particles. By exploiting these correlations, quantum key distribution (QKD) offers security fundamentally superior to classical cryptography. The security of QKD is intrinsically linked to the violation of Bell inequalities, ensuring that any eavesdropping attempt will be detectable. The non-classical features of entanglement are essential for these advancements.

Moreover, understanding Bell inequalities is vital for developing and evaluating quantum technologies in various contexts. Quantum computing, quantum sensing, and quantum communication all leverage the intrinsically non-classical features of entanglement, features confirmed by the violation of Bell inequalities. As we design and implement these technologies, we need a robust understanding of the theoretical underpinnings, which includes a comprehensive grasp of the experimental verification and implications of Bell's theorem. It is not merely an abstract theoretical concept; it forms a cornerstone of modern quantum technologies.

The experimental methodologies used to test Bell inequalities are sophisticated and varied. They generally involve creating entangled pairs of particles, often photons, measuring their properties along different directions, and analyzing the statistical correlations of the measurement outcomes. The analysis hinges on meticulously comparing the experimental results to the predictions made by both local realism and quantum mechanics. Discrepancies with the Bell inequalities' constraints indicate a violation of local realism. Experimental success in this area has spurred further exploration and development in quantum technologies.

It's also important to note the ongoing debates and refinements in the interpretations and testing of Bell inequalities. Loopholes, such as detection loopholes and locality loopholes, have been identified and addressed through increasingly refined experimental designs. The ongoing research seeks to close all possible loopholes, providing ever-stronger evidence for the violation of local realism and affirming the validity of quantum mechanics' predictions. This ongoing work contributes significantly to improving our understanding of the fundamental principles of the quantum world and its implications for various technologies. The exploration and improvement of the tests themselves are as important as the results.

In summary, Bell inequalities provide a rigorous and experimentally testable framework for distinguishing between local realism and quantum mechanics. Their violation, robustly demonstrated in numerous experiments, confirms that quantum mechanics indeed operates outside the constraints of classical intuition. This has profound implications for our understanding of the quantum world and underpins the development of various quantum technologies. As we move further into the quantum age, a deep comprehension of Bell inequalities remains crucial for both theoretical physicists and engineers involved in quantum technologies. Furthermore, this area of physics is continuously evolving with new theoretical challenges and technological improvements.

5 Advanced NMR Techniques

Solid-state NMR differs significantly from solution-state NMR, primarily due to the restricted molecular motion in solid samples. In solution, rapid tumbling averages out anisotropic interactions, simplifying the spectra. Solids, however, lack this averaging, leading to broad, complex lineshapes resulting from interactions such as chemical shift anisotropy (CSA), dipolar couplings, and quadrupolar interactions (for nuclei with spin > 1/2). Advanced techniques like magic angle spinning (MAS) are crucial to overcome these challenges. MAS involves spinning the sample at high speeds around an axis tilted at the "magic angle" (54.74°) relative to the magnetic field. This significantly reduces the CSA and dipolar broadening, resulting in much narrower lines and improved spectral resolution. Other sophisticated pulse sequences, such as cross-polarization (CP) and multiple-quantum magic angle spinning (MQMAS), enhance signal sensitivity and resolve overlapping resonances in complex solid-state spectra. These techniques are essential for investigating the structure and dynamics of materials such as polymers, catalysts, and biological tissues. The interpretation of solid-state NMR spectra often requires sophisticated simulations and modeling to extract detailed structural information.

Two-dimensional (2D) NMR spectroscopy represents a substantial advancement over one-dimensional (1D) NMR. 1D NMR provides information on chemical shifts, but often suffers from spectral overlap, making it difficult to fully resolve complex molecules. 2D NMR overcomes this limitation by spreading the information across two frequency dimensions. This is achieved through carefully designed pulse sequences that introduce correlations between different nuclei or between different aspects of a single nucleus's interactions. Several fundamental 2D NMR experiments exist, each providing unique insights. Correlation spectroscopy (COSY) reveals through-bond connectivities between protons, allowing for the determination of spin coupling networks within molecules. Nuclear Overhauser effect spectroscopy (NOESY) reveals through-space proximity between protons, providing crucial information about molecular conformation and dynamics. Heteronuclear correlation spectroscopy (HSQC and HMBC) techniques are invaluable for establishing correlations between different types of nuclei, such as 1H and 13C, leading to complete assignments in complex molecules. The power of 2D NMR lies in its ability to unravel intricate molecular architectures.

Advanced pulse sequences play a critical role in modern NMR. These sophisticated sequences, often involving multiple radio frequency pulses and carefully controlled delays, are designed to achieve specific goals, such as enhancing signal-to-noise ratios, suppressing unwanted signals, or generating specific types of coherence. For example, techniques like pulsed field gradients (PFGs) are widely employed for selective excitation and coherence pathway selection, crucial for improving the resolution and sensitivity of many experiments. Other advanced pulse sequences, such as those used in diffusion-ordered NMR spectroscopy (DOSY), provide information about molecular size and shape. DOSY allows for the separation of components in complex mixtures based on their translational diffusion coefficients. These advanced pulse techniques often require precise control of the radio frequency pulses and gradients and advanced data processing. The development and refinement of pulse sequences remain an active area of research, constantly pushing the boundaries of NMR capabilities.

Understanding and correctly interpreting advanced NMR data requires expertise in both experimental techniques and data analysis. Signal processing steps often include Fourier transformation, spectral referencing, baseline correction, and phasing. These steps are crucial for obtaining high-quality spectra. In addition, advanced methods such as spectral deconvolution, quantitation, and line-shape analysis are used to extract accurate and precise information. More complex data, such as that from 2D NMR, often necessitate specialized software and algorithms for data processing and interpretation. Software packages such as NMRPipe and Topspin allow for complex data analysis and spectral manipulation. The mastery of these data processing techniques is essential for obtaining meaningful results from sophisticated NMR experiments. Careful attention to detail at every step, from sample preparation to data analysis, is critical for reliable interpretations.

The future of NMR continues to evolve rapidly. New developments in magnet technology, probe design, and pulse sequences are expanding the capabilities of this technique at an unprecedented rate. Higher magnetic field strengths provide improved resolution and sensitivity, allowing for the study of increasingly complex systems. Advanced probe designs, such as cryoprobes, further enhance sensitivity by reducing electronic noise. The development of new pulse sequences and data analysis methods is critical for unlocking additional information from NMR data. Furthermore, advances in computational power allow for simulations and theoretical modeling of NMR experiments, helping to bridge the gap between experimental observations and molecular structure and dynamics. The combination of these technological advances and innovative theoretical developments promises to unlock new frontiers in our understanding of a broad range of scientific systems, expanding the already profound impact of NMR in diverse fields.

5.1 Solid-State NMR

Solid-state NMR presents unique challenges compared to solution-state NMR. In liquids, the rapid tumbling of molecules averages out anisotropic interactions, simplifying spectra. Solids, however, lack this averaging, leading to broad and complex spectra reflecting the diverse orientations of molecules within the sample. This broadening obscures much of the detailed information one would typically extract from solution-state NMR, making specialized techniques crucial for unraveling structural and dynamical properties.

Understanding these anisotropic interactions is fundamental. The chemical shift, a key parameter in NMR, becomes a tensor in solids, not a single scalar value as in solution. This tensorial nature arises because the shielding of the nucleus by surrounding electrons is direction-dependent. Similarly, the dipolar coupling between nuclear spins, a significant source of line broadening, depends strongly on the internuclear vector's orientation relative to the magnetic field. Quadrupolar interactions, present for nuclei with spin greater than 1/2, add another layer of complexity due to the interaction of the nuclear quadrupole moment with the electric field gradient at the nucleus.

Magic angle spinning (MAS) is a cornerstone technique in solid-state NMR. By spinning the sample at a specific angle (the magic angle, approximately 54.7 degrees) relative to the magnetic field, the anisotropic interactions are partially averaged, leading to narrower lines. The spinning frequency determines the effectiveness of this averaging; higher spinning speeds are needed to average out stronger interactions. Advanced MAS techniques, such as double-rotation (DOR) or dynamic angle spinning (DAS), can further improve spectral resolution by averaging out additional anisotropic interactions.

Beyond MAS, other techniques are crucial for extracting information from solid-state NMR spectra. Cross-polarization (CP) enhances the signal from low-sensitivity nuclei (like ^{13}C) by transferring magnetization from a more sensitive nucleus (like ^{1}H). This technique is particularly valuable in materials containing low concentrations of the target nucleus. Decoupling techniques suppress the broadening effects of heteronuclear dipolar couplings, thereby simplifying spectra and improving resolution. Multiple-pulse sequences are applied to average out homonuclear dipolar couplings, enhancing spectral resolution.

Advanced solid-state NMR experiments often involve multiple pulses, delays, and rotations. These sequences are designed to manipulate the spin system in specific ways, allowing for the selective observation of particular interactions or the separation of overlapping resonances. Techniques like two-dimensional (2D) solid-state NMR spectroscopy, discussed in the next chapter, offer additional dimensions for spectral analysis, enabling researchers to unravel complex spin networks and reveal structural information not accessible through one-dimensional experiments. They also provide valuable insights into the dynamics of the molecules within the solid-state, complementing structural information.

Analyzing solid-state NMR spectra often involves sophisticated data processing and spectral simulation techniques. Spectral deconvolution methods are crucial for resolving overlapping peaks, allowing for the extraction of precise chemical shift and coupling parameters. Simulations, based on models of molecular structure and dynamics, are then used to compare experimental data to theoretical predictions and refine understanding. These techniques often rely on advanced algorithms and computational resources, highlighting the intersection of NMR with other fields like computational chemistry.

The practical applications of solid-state NMR are wide-ranging. It is an invaluable tool in material science, providing detailed information about the structure, dynamics, and interactions of molecules in various materials, from polymers and ceramics to biomolecules and catalysts. It provides critical information about the structure and interactions of molecules in solids, allowing scientists to explore complex systems with unprecedented detail. In fact, this technique enables scientists to unveil important structural and functional aspects of biological macromolecules in their natural state, significantly furthering our understanding of life processes.

Solid-state NMR has evolved into a powerful tool, enabling researchers to study a vast range of materials and systems. While the basic principles are rooted in the fundamental interactions of nuclear spins with magnetic fields, the experimental techniques and data analysis methods have become increasingly sophisticated. Understanding these techniques and their underlying principles is essential for anyone seeking to leverage the power of solid-state NMR in their research. The continued advancements in this field promise even more detailed structural and dynamic insights into the fascinating world of solids in the years to come.

5.2 2D NMR Spectroscopy

Two-dimensional (2D) NMR spectroscopy significantly enhances the capabilities of basic NMR by providing more detailed information about molecular structure and dynamics. Unlike one-dimensional NMR, which displays signals along a single frequency axis, 2D NMR spreads the information across two frequency axes, dramatically improving spectral resolution and allowing us to observe interactions not readily apparent in 1D experiments. This increased resolution is crucial for analyzing complex molecules, where spectral peaks often overlap in 1D spectra, obscuring important information.

Consider a scenario involving a complex organic molecule with numerous protons. In a standard 1D NMR spectrum, the signals from these protons may overlap, making it challenging to distinguish individual resonances and assign them to specific atoms within the molecule. However, a 2D NMR experiment, such as COSY (Correlation Spectroscopy), can resolve this issue. COSY uses a series of pulses and time delays to create a spectrum with two frequency axes, revealing correlations between protons that are coupled to one another through bonds. Protons on adjacent carbons, for example, will show a cross-peak in the 2D spectrum, providing a clear map of the molecule's connectivity. The power of 2D NMR lies in its ability to reveal these correlations, providing structural information beyond the reach of 1D techniques. A variety of 2D NMR experiments exists, each designed to probe specific types of interactions. For instance, NOESY (Nuclear Overhauser Effect Spectroscopy) detects through-space interactions between protons that are close in proximity, even if they are not directly bonded. This is extremely useful for determining the three-dimensional structure of molecules, particularly proteins and other biomolecules. Imagine trying to unravel the intricate folding pattern of a protein using only 1D NMR; it would be a near-impossible task. NOESY provides crucial distance constraints, simplifying the process considerably.

To understand the practical implementation, imagine you're working with a newly synthesized compound. A 1D NMR spectrum provides basic information about the types and numbers of protons and carbons present. However, to determine the exact connectivity and three-dimensional structure, you would need to employ 2D techniques. The choice of specific 2D experiment depends on the information you seek. If connectivity is the primary goal, COSY or TOCSY (Total Correlation Spectroscopy) would be suitable choices. For determining three-dimensional structure, NOESY or ROESY (Rotating-frame Overhauser Enhancement Spectroscopy) are preferable.

The data obtained from 2D NMR experiments are presented as contour plots, with each peak representing a correlation between two nuclear spins. The x and y axes correspond to the chemical shifts of the two interacting nuclei. The intensity of the peak reflects the strength of the interaction. Analyzing these 2D spectra requires some level of expertise, but specialized software packages are available to aid in peak identification and assignment. These programs often incorporate advanced algorithms for automated peak picking and integration, streamlining data analysis. Moreover, the advancement in pulse sequences and experimental parameters optimization allows for more efficient data acquisition, reducing experiment times significantly.

Understanding the underlying principles of 2D NMR, the different experiment types, and data analysis techniques is essential for obtaining meaningful results. While sophisticated software packages significantly simplify the analysis, a firm grasp of the theoretical foundation is crucial for interpreting the data correctly. This ensures that the information obtained accurately reflects the molecular structure and dynamics under investigation. The insights gleaned from 2D NMR are pivotal for diverse applications, ranging from the structural elucidation of small molecules and complex natural products to characterizing interactions within proteins and other large biomolecules. In modern chemistry and biochemistry research, 2D NMR is an indispensable tool.

Furthermore, the development of advanced pulse sequences and sophisticated data processing methods continues to expand the capabilities of 2D NMR spectroscopy. Techniques such as heteronuclear correlation experiments (e.g., HSQC, HMBC) provide correlations between different types of nuclei (e.g., 1H and 13C), yielding detailed insights into molecular structure and connectivity. These advancements enable researchers to address increasingly complex questions in various scientific fields, solidifying the central role of 2D NMR in modern scientific research. The combination of theoretical understanding and practical expertise in experimental design and data analysis is key to successfully utilizing this powerful technique.

In conclusion, 2D NMR spectroscopy offers a powerful and indispensable tool for advancing our understanding of molecular structure and dynamics. Its ability to resolve overlapping signals and reveal subtle interactions, inaccessible through 1D NMR, is transforming our approach to complex chemical and biological systems. The continuous development of new pulse sequences, increased magnetic field strengths, and improved data processing capabilities further enhances the capabilities of this technique, solidifying its role as a cornerstone of modern NMR spectroscopy. By understanding its principles, experimental parameters, and data analysis strategies, researchers can leverage the power of 2D NMR to unlock new insights into the intricacies of the molecular world.

6 Applications of EPR and NMR

The power of NMR and EPR extends far beyond the fundamental principles already discussed. These techniques are indispensable tools across diverse scientific fields, impacting our understanding of the world at a molecular level and even contributing to life-saving medical technologies. Their applications are so vast they are shaping numerous sectors.

Medical imaging, specifically Magnetic Resonance Imaging (MRI), relies entirely on NMR principles. MRI uses strong magnetic fields and radio waves to create detailed images of the human body's internal structures. The differing resonance frequencies of protons in various tissues allow for high-resolution contrast, allowing physicians to diagnose diseases and injuries with unprecedented accuracy. Contrast agents, specifically designed molecules that alter the relaxation times of nearby protons, further enhance the image quality and specificity, assisting in the detection of tumors or vascular abnormalities. This non-invasive technique has revolutionized medical diagnostics, enabling earlier and more precise interventions.

Beyond medicine, material science heavily leverages both EPR and NMR. EPR is exceptionally sensitive to unpaired electrons, making it invaluable for characterizing defects, impurities, and oxidation states within materials. Researchers utilize EPR to study semiconductors, polymers, and catalysts, gaining insights into their electronic structures and behaviors. For example, understanding the behavior of paramagnetic centers in newly synthesized materials is critical for developing novel technologies like more efficient solar cells or more advanced batteries. This has implications from improving material durability to boosting energy efficiency.

NMR, on the other hand, provides detailed information on molecular structure and dynamics in materials. Solid-state NMR, in particular, offers a unique window into the structure and interactions within solid materials, providing crucial data on crystal structures, polymer chain conformations, and the dynamics of molecules confined within porous materials. This understanding is critical for designing new materials with specific properties, such as advanced polymers with enhanced strength and flexibility or novel porous materials for efficient gas storage or catalysis. The data obtained isn't merely descriptive; it allows for predictive modeling and informed synthesis.

Chemical analysis benefits significantly from both techniques. NMR spectroscopy serves as a primary tool for identifying and quantifying molecules in complex mixtures. Its ability to distinguish different chemical environments within molecules, combined with advanced techniques like 2D NMR, provides a detailed fingerprint of the molecular composition, crucial in various fields including pharmaceutical drug discovery, environmental monitoring, and food science. EPR spectroscopy complements NMR by providing insights into reaction mechanisms involving free radicals, providing invaluable information for chemists studying reaction dynamics and mechanisms, especially those involving oxidation or reduction.

EPR and NMR are not merely analytical tools; they also play crucial roles in studying chemical kinetics and reaction mechanisms. By tracking the changes in the EPR or NMR signals over time, researchers can gain detailed insights into the rates and pathways of chemical reactions. This is especially important in understanding complex chemical processes like catalysis and polymerization. Understanding kinetics, in turn, allows for the optimization of reaction conditions and the design of more efficient processes, further driving innovation in chemical technology.

In summary, the applications of Electron Paramagnetic Resonance (EPR) and Nuclear Magnetic Resonance (NMR) are incredibly wide-ranging and impactful. From medical diagnostics to material science and chemical analysis, these techniques provide indispensable tools for researchers across various fields. The ongoing development and refinement of these techniques promise even more significant breakthroughs in the future, unlocking further understanding and technological advancements across various scientific domains and beyond. Their importance is underscored by their continued development and refinement, promising yet more impactful contributions to science and technology. Their combined power allows scientists to probe the intricacies of the molecular world with unprecedented precision.

6.1 Medical Imaging

Magnetic Resonance Imaging (MRI), a cornerstone of modern medical diagnostics, leverages the principles of Nuclear Magnetic Resonance (NMR) to create detailed images of the human body. Unlike X-rays or CT scans that rely on ionizing radiation, MRI uses powerful magnets and radio waves, making it a remarkably safe and versatile imaging technique. Its non-invasive nature allows for repeated scans, crucial for monitoring disease progression or treatment response. The process begins by placing the patient within a strong magnetic field, aligning the nuclear spins of hydrogen atoms, abundant in water and fat throughout the body. Then, radiofrequency pulses are applied, temporarily disrupting this alignment, and the subsequent relaxation process—the return to equilibrium—is meticulously monitored. The variations in relaxation times, unique to different tissues and their molecular environments, provide the fundamental contrast that allows for the exquisite anatomical detail visible in MRI scans.

This exquisite detail arises from the sophisticated signal processing employed in MRI. The radio waves emitted during relaxation are incredibly weak, requiring highly sensitive detection techniques and complex algorithms to reconstruct the image. Different MRI sequences—variations in the radiofrequency pulses and magnetic field gradients—are carefully tailored to highlight specific tissue characteristics. For instance, T1-weighted images emphasize differences in relaxation times related to the longitudinal magnetization recovery, revealing anatomical structures with high spatial resolution. T2-weighted images, sensitive to transverse magnetization decay, excel at highlighting the presence of edema or lesions, areas characterized by slower relaxation processes. Furthermore, diffusion-weighted imaging, exploiting the movement of water molecules within the tissues, can differentiate between grey and white matter in the brain, detect stroke, and even assess the viability of transplanted organs.

Beyond anatomical imaging, functional MRI (fMRI) has revolutionized neuroscience. fMRI relies on the blood-oxygen-level-dependent (BOLD) contrast, indirectly measuring neural activity by detecting changes in blood oxygenation. Active brain regions consume more oxygen, leading to alterations in the magnetic susceptibility of hemoglobin, the oxygen-carrying protein in red blood cells. This subtle change in the magnetic environment is detected by fMRI, generating images showing brain activity during various tasks, allowing researchers to map cognitive processes, emotional responses, and the effects of disease. The ability to monitor brain activity in real-time holds immense potential for understanding neurodegenerative diseases like Alzheimer's and Parkinson's, optimizing treatment strategies, and developing innovative neurorehabilitation approaches.

However, MRI is not without limitations. The strong magnetic fields can pose safety risks for individuals with certain metallic implants. The scan time can be lengthy, requiring patients to remain motionless for extended periods, which can be challenging for some individuals. Furthermore, MRI is expensive compared to other imaging techniques. Despite these limitations, MRI remains a cornerstone of medical imaging, providing unparalleled anatomical detail and functional insights. Advancements in MRI technology, including improved gradient coils, faster acquisition sequences, and advanced image reconstruction algorithms, constantly enhance its capabilities, pushing the boundaries of clinical applications.

The development of new contrast agents continues to broaden the scope of MRI. These agents, selectively targeting specific tissues or cellular processes, enhance image contrast and allow for more precise diagnosis. For example, gadolinium-based contrast agents enhance the visualization of tumors, while superparamagnetic iron oxide nanoparticles can target specific cells or tissues. The ongoing research in this field aims to develop even more targeted and sensitive contrast agents, tailored to specific diseases and biological processes, further optimizing the effectiveness and precision of MRI. The integration of MRI with other imaging modalities, such as PET or SPECT, is also an area of active research, aiming to combine the anatomical detail of MRI with the metabolic information provided by nuclear medicine techniques. This synergistic approach promises to provide even more comprehensive insights into disease processes, opening avenues for improved diagnosis and more effective therapies. This multi-modal approach is particularly powerful for cancer detection, where combining anatomical and functional data offers a much clearer picture of the tumor's extent, its cellular characteristics, and its metabolic activity.

6.2 Material Science

Material science significantly benefits from NMR and EPR techniques, offering unparalleled insights into the structure and dynamics of various materials. Understanding the interactions of nuclear and electron spins with their surroundings provides crucial information about material properties, unattainable through other methods. For example, NMR can reveal the local environment of specific atoms within a solid, such as the presence of impurities or defects, impacting the material's overall performance. This detailed microscopic analysis is invaluable for optimizing material synthesis and processing, leading to improvements in material characteristics.

Solid-state NMR, a specialized technique, is particularly well-suited for analyzing the structure and dynamics of solid materials, from crystalline structures to amorphous solids and polymers. It directly probes the local environment of atoms, revealing information about their bonding, mobility, and interactions. For instance, studying the diffusion of atoms within a solid, crucial in processes like sintering or doping, relies heavily on solid-state NMR. The ability to examine these processes non-destructively makes it a powerful tool. In contrast to other methods which might only provide average structural information, solid-state NMR offers atomic-level detail, uncovering subtle differences that directly affect macroscopic properties. Consider the design of new alloys - NMR allows researchers to pinpoint the effects of compositional changes and their resulting impact on the material's strength, ductility and electrical conductivity.

EPR, focusing on unpaired electrons, proves equally valuable. Many materials exhibit paramagnetic properties, arising from unpaired electrons in their structure. EPR spectroscopy measures the interaction of these unpaired electrons with their environment, providing critical information about the material's electronic structure, defects, and oxidation states. It is particularly useful in characterizing semiconductors and catalysts where understanding the behavior of unpaired electrons is essential. For example, in catalysis research, EPR can identify the presence of active sites, enabling researchers to tune the catalyst's performance for specific reactions by tailoring the material properties and understanding the role of these unpaired electrons.

Furthermore, combining NMR and EPR with other techniques like X-ray diffraction or electron microscopy provides a comprehensive understanding of a material's structure and properties. These combined approaches allow for a holistic characterization, correlating microscopic information with macroscopic behavior. Imagine the development of new high-temperature superconductors; integrating EPR's insights into the electronic structure with NMR's understanding of atomic-scale dynamics helps optimize synthesis strategies and predict material performance at various temperatures and pressures. This multi-faceted approach is vital for developing materials with superior properties tailored to specific applications.

The non-destructive nature of both NMR and EPR is a significant advantage. They offer a way to study materials without altering their properties, essential for analyzing delicate or valuable samples. Researchers can monitor changes in material properties in situ and real time during processing or exposure to specific conditions, offering valuable information on material stability and degradation pathways. This dynamic analysis, unattainable by other destructive testing methods, allows researchers to refine processing parameters and design more durable and reliable materials.

The sensitivity of NMR and EPR to small changes in local environments enables the identification of subtle structural differences that might be missed by other techniques. For example, the detection of minute changes in crystal structure or the presence of low concentrations of impurities becomes readily achievable, providing crucial insights into how these features affect overall material performance. In the realm of material degradation, for instance, NMR can precisely track the subtle changes in molecular structure and polymer chains during degradation processes, leading to a better understanding of the degradation mechanism and improvements in materials design to enhance stability and longevity. Such precise measurements are essential in ensuring the reliability and performance of materials in various applications.

Finally, the versatility of these techniques allows their adaptation to a wide array of materials, from inorganic solids and polymers to biological tissues and porous materials. Researchers can tailor experimental parameters to obtain specific information relevant to their material systems and address unique challenges. For example, using various NMR pulse sequences or EPR spectral parameters, one can extract detailed information on both the static and dynamic aspects of a material's properties, including molecular mobility, chemical exchange, and electronic structure, with high precision and accuracy. This flexibility and versatility make NMR and EPR indispensable tools in contemporary material science research and development.

6.3 Chemical Analysis

NMR and EPR spectroscopies offer powerful tools for chemical analysis, going far beyond simple identification. Their ability to provide detailed structural information, including bonding arrangements, molecular dynamics, and even quantitative data on component concentrations,

makes them indispensable in numerous fields. For beginners, grasping the fundamental principles is key to harnessing their full potential.
Understanding the chemical shift, a fundamental concept in both NMR and EPR, is crucial. This phenomenon describes the variation in resonance frequency of a nucleus or electron due to its local electronic environment. The electron density surrounding an atom affects the magnetic field experienced by the nucleus, causing shifts in its resonance frequency. These shifts are highly sensitive to the chemical environment, providing valuable insights into molecular structure. For instance, a carbon atom in a carbonyl group will have a drastically different chemical shift compared to a carbon in a methyl group. This allows for the differentiation of various functional groups within a molecule. Analyzing the chemical shift patterns in a spectrum gives chemists a fingerprint of the molecule.
Furthermore, coupling constants play a critical role in determining molecular connectivity. These constants represent the interaction between the nuclear or electron spins of neighboring atoms. The magnitude of this interaction, expressed in Hertz, provides information about the number of nearby atoms and the bonds connecting them. For instance, a large coupling constant indicates a strong interaction, often observed between nuclei directly bonded to each other. Conversely, a small coupling constant signifies a weaker interaction, suggesting more distant atoms. Analyzing these coupling patterns is crucial for piecing together the complete connectivity map of a molecule.
Beyond simple structural elucidation, NMR and EPR techniques offer quantitative capabilities. The signal intensity in a spectrum is directly proportional to the concentration of the molecule being analyzed. This

allows researchers to quantitatively determine the amounts of different compounds in a mixture. Careful calibration, proper signal processing and an understanding of relaxation mechanisms are essential for obtaining accurate quantitative data. Indeed, the relative peak intensities in an NMR spectrum directly reflect the relative abundance of different nuclei in the molecule. This simple principle allows one to readily determine the stoichiometry of a molecule, which is crucial in confirming or rejecting proposed structures.

Furthermore, advanced techniques like 2D NMR spectroscopy provide a level of detail that surpasses what simpler approaches can offer. These methods correlate signals from different nuclei or atoms, revealing through-space or through-bond connectivity patterns beyond the capabilities of one-dimensional spectroscopy. This added dimensionality enables the comprehensive mapping of complex molecules, especially those with many atoms and structural isomers. Understanding these advanced techniques opens up a vast toolbox for even the most intricate molecular analysis challenges. Beginners often find the initial hurdle to be the vastness of this information, but a systematic approach, working incrementally from basic principles to these more complex techniques, will build a firm foundation in this critical area of chemical analysis.

Finally, remember that sample preparation is paramount to obtaining high-quality NMR and EPR data. Factors such as solvent selection, concentration, temperature, and the presence of paramagnetic impurities can all significantly affect the spectra obtained. Careful consideration and optimization of these experimental conditions are essential in ensuring reliable and accurate results. The choice of solvent, for instance, can influence the chemical shifts and coupling constants observed, requiring an understanding of solvent effects on NMR and EPR spectroscopy. This might involve choosing a deuterated solvent to avoid interference from proton signals in proton NMR experiments. Such attention to experimental detail underscores the importance of a holistic approach to chemical analysis, encompassing not only spectral interpretation but also careful sample preparation and experimental design. Successful application of these techniques, then, requires diligence and a thorough understanding of both the theoretical underpinnings and the practical aspects of these powerful tools.

7 Future Directions in Quantum Spin Resonance

Quantum sensing represents a significant frontier. Harnessing the exquisite sensitivity of quantum systems, particularly entangled spins, promises unprecedented precision in measurement. Imagine detectors capable of sensing minute changes in magnetic fields, enabling advancements in medical diagnostics, materials characterization, and fundamental physics research. Developing robust and scalable quantum sensors, however, requires overcoming significant technological hurdles, including minimizing decoherence effects and improving control over individual quantum systems. This involves innovative designs, utilizing materials with extended coherence times, and developing more efficient control protocols. Further research is needed in developing novel sensing modalities and algorithms to extract meaningful information from the quantum sensor output. Exploring the possibilities with nitrogen-vacancy centers in diamond, for instance, offers a promising route. The path forward demands rigorous experimental investigation and theoretical modeling.

The quest for advanced quantum control methods is paramount. Precise manipulation of individual quantum spins is essential for realizing the full potential of quantum technologies. Sophisticated pulse sequences, optimized for specific tasks, are already being implemented, but achieving even greater control over the quantum evolution necessitates the development of more advanced techniques. This includes exploring novel control mechanisms based on advanced laser technologies, microwave manipulation, and potentially even employing hybrid quantum systems. For instance, combining optical control with microwave pulses may open pathways to a higher level of control. The challenge lies in minimizing errors and improving the fidelity of quantum control operations, thus increasing the precision in preparing specific quantum states. These improvements are crucial for building more complex quantum devices with higher functionality.

Discovering novel materials with superior spin properties is key. Current technologies often rely on specific materials with limited performance. The search for new materials with enhanced coherence times, higher spin polarization, and improved scalability is a critical area of ongoing research. This involves exploring a wide range of materials, including novel two-dimensional materials, topological insulators, and hybrid organic-inorganic systems. The development of sophisticated material synthesis techniques and advanced characterization methods is essential to identify and refine these materials, especially those with enhanced spin-spin interactions which can be used in quantum computing and other applications. Such a discovery could revolutionize the field, enabling the creation of more efficient and powerful quantum spin devices.

Theoretical developments underpin future advancements. A deeper understanding of the fundamental quantum dynamics governing spin interactions is needed to design better quantum devices and algorithms. This requires the development of improved theoretical models that accurately account for complex interactions, including decoherence and environmental noise. Addressing the challenges posed by the many-body problem, improving theoretical methods for simulating large quantum systems, and developing new approaches to quantify entanglement are all crucial avenues of future research. Theoretical work should be closely coupled with experimental work, guiding the design of new experiments and interpreting experimental results within a rigorous theoretical framework.

Technological advancements are inseparable from progress in the field. Developments in laser and microwave technology, cryogenics, and nanofabrication are crucial for building better quantum devices. Improved detectors and signal processing techniques are needed to overcome the challenges of detecting weak quantum signals. Developing scalable and robust manufacturing processes for quantum devices is equally critical for translating research findings into practical applications. The convergence of these technological advancements with theoretical and materials research will drive the future development of this exciting area. Microfluidic technologies, for example, might bring a new dimension to the process and enable the mass production of quantum devices, reducing the costs and bringing the technology to a wider market.

7.1 Quantum Sensing

Quantum sensing harnesses the exquisite sensitivity of quantum systems to detect minute changes in their environment. Unlike classical sensors, which rely on the average behavior of many particles, quantum sensors exploit quantum phenomena like superposition and entanglement to achieve unprecedented precision. Imagine a tiny compass needle, exquisitely sensitive to even the faintest magnetic field fluctuations - this is the essence of a quantum sensor. The inherent quantum properties amplify the signal, leading to detection capabilities far surpassing classical methods.

This remarkable sensitivity stems from several key quantum features. Superposition, for instance, allows a quantum system to exist in multiple states simultaneously, increasing its responsiveness to external stimuli. A quantum sensor in superposition can simultaneously "feel" various aspects of its environment, providing a more comprehensive picture than a classical counterpart confined to a single state. The exploitation of entangled states takes this further, linking the fates of multiple quantum systems. Measuring one provides instantaneous information about the others, offering a dramatic improvement in signal-to-noise ratio. These entangled particles behave as a collective, enhancing the measurement precision beyond what individual particles could achieve.

Consider the application of nitrogen-vacancy (NV) centers in diamond as quantum sensors. These NV centers, atomic-scale defects in the diamond lattice, possess a unique electron spin that exhibits exceptional coherence properties, making them ideally suited for sensing various physical parameters. Their response to subtle magnetic field changes or temperature fluctuations allows for high-resolution measurements. Scientists have employed NV centers for magnetic field imaging with nanometer spatial resolution, opening up exciting possibilities for nanoscale science and biomedical applications.

Beyond magnetic field sensing, quantum sensors find use in a multitude of applications. They have shown great promise in gravimetry, measuring variations in gravitational acceleration with unparalleled accuracy. Such precision is pivotal in geophysical surveys, exploring the Earth's internal structure, and even detecting subtle changes related to seismic activity. The potential also extends to detecting minute changes in electric fields, offering improved sensors for materials science or even biological systems, capable of detecting the electrical signals associated with neuronal activity. In addition, quantum sensors based on trapped ions or superconducting circuits offer alternative approaches with potentially even greater sensitivity and controllability. The continuous development of novel quantum materials and improved experimental techniques promises to further enhance these capabilities.

Developing quantum sensors involves careful design and control of the quantum system. Precise control over the interactions between the quantum system and its environment is crucial for accurate measurements. This often requires advanced laser techniques, microwave pulses or other methods for manipulating the quantum states with high fidelity. Minimizing the influence of environmental noise is paramount as this noise can easily mask the subtle signals being measured. This requires sophisticated shielding and cooling techniques, and careful selection of materials to minimize the effects of unwanted interactions.

Data analysis also plays a crucial role, as sophisticated algorithms are often needed to extract meaningful signals from noisy data.

However, building and operating quantum sensors present unique challenges. The delicate nature of quantum systems makes them highly susceptible to environmental noise and decoherence. This loss of coherence limits the sensor's sensitivity and measurement time. Overcoming these issues requires sophisticated techniques for isolating the quantum system from its environment, such as using cryogenic cooling or vacuum chambers, and developing advanced error correction codes to mitigate the effects of noise and decoherence. The complexity of the experimental setup and the need for specialized equipment also pose significant hurdles to widespread adoption. Nevertheless, active research in these areas continues to push the boundaries of what's achievable, leading to increasingly robust and sensitive quantum sensors.

The future of quantum sensing looks incredibly bright. Advancements in material science and quantum control techniques will likely lead to even more sensitive and versatile sensors. The development of integrated quantum sensors promises to shrink the size and cost of these devices, leading to their integration into a broad range of applications. Exploring novel quantum systems, such as topological qubits, could further enhance sensor stability and performance, leading to a new generation of quantum sensors. This promises advancements in various fields, from medical diagnostics and fundamental research to environmental monitoring and industrial applications. The journey towards practical, widespread implementation of quantum sensors is ongoing, promising a future of remarkably precise and powerful measurement capabilities.

This evolution is underpinned by continuous innovations in the fields of quantum control, materials science, and data analysis. Ultimately, the impact of quantum sensing on scientific discovery and technological advancement is expected to be truly transformative.

7.2 Advanced Quantum Control

Precise control over quantum systems is paramount for realizing the full potential of quantum technologies. This requires sophisticated techniques that go beyond basic manipulation. We need to move beyond simple pulses and explore more intricate control schemes. These advanced techniques are crucial for tasks like creating specific quantum states, implementing complex quantum algorithms, and enhancing the sensitivity of quantum sensors.

Achieving high-fidelity quantum gates is a major challenge. Imperfect control leads to errors, significantly impacting the performance of quantum computers. Advanced techniques such as dynamical decoupling, which mitigate the effects of noise, are essential. Implementing optimal control theory allows for designing pulse sequences that minimize errors and maximize the fidelity of quantum operations. This involves sophisticated algorithms that optimize control parameters to achieve a desired quantum evolution.

Furthermore, the precision of control extends beyond simply creating specific states; it also dictates the ability to maintain coherence. Quantum coherence, the fragile superposition of quantum states, is susceptible to decoherence from environmental interactions. Active control techniques, such as quantum error correction codes, aim to actively protect quantum information from these disruptive influences, extending coherence times and enhancing the reliability of quantum computations. Think of it like carefully shielding a delicate instrument from external vibrations - maintaining coherence requires similar precision and care.

Beyond gate fidelity, advanced control also impacts the scalability of quantum technologies. Controlling a small number of qubits is one thing; scaling up to hundreds or thousands presents a significantly more complex challenge. Efficient control strategies are essential for managing the interactions between many qubits, minimizing crosstalk, and effectively manipulating the overall system. This necessitates a shift from individual qubit control to collective control schemes.

Another crucial aspect is the development of novel control methodologies. Research into optimal control theory, quantum optimal control, and other advanced control algorithms is essential for designing precise and efficient control schemes. These theoretical advancements translate directly into practical improvements in experimental techniques. For example, the use of machine learning algorithms for optimizing control parameters is showing significant promise. Adaptive control strategies, which constantly monitor and adjust control based on system feedback, are also being developed to counteract unforeseen disturbances and enhance robustness.

Finally, consider the development of new materials and experimental platforms specifically designed for optimal quantum control. The choice of physical system greatly influences the achievable level of control. Superconducting circuits, trapped ions, and neutral atoms each have their own advantages and challenges concerning control. Careful material selection, optimized device design, and innovative experimental setups are crucial for pushing the boundaries of what is possible in quantum control. The quest for better materials and improved experimental techniques forms a continuous cycle of advancement, constantly feeding into improved control methods. This continuous development loop represents the future of the field.

7.3 New Materials

The quest for novel materials with exceptional properties for quantum spin resonance applications is a driving force in modern materials science. Imagine materials exhibiting unprecedented levels of coherence, allowing for the manipulation and control of quantum states over extended periods. This is the realm we are entering, a space where manipulating electron and nuclear spins for advanced applications like quantum computing becomes significantly more feasible. We need materials that are exceptionally pure, with minimal defects that could disrupt spin coherence. Think of it as striving for perfection at the atomic level, removing any impurities or imperfections that could hinder the quantum phenomena we're trying to exploit. This necessitates advanced synthesis techniques, pushing the boundaries of chemical and physical methods to create materials with previously unattainable purity and structural control.

Developing materials with tailored electronic and magnetic properties is paramount. We need materials where the spin states are easily controllable, where we can precisely tune the interactions between spins, and where the environment is optimized to protect the delicate quantum coherence. This involves a deep understanding of the interplay between structure, chemistry, and the resulting magnetic properties. This detailed understanding is key in predicting and designing materials with specific spin characteristics, essential for advancing quantum spin resonance technologies. For example, we could design materials that enhance the signal strength of NMR or EPR experiments, leading to more sensitive and precise measurements. Or perhaps we need materials that reduce decoherence, leading to longer spin lifetimes.

Beyond the pursuit of idealized purity, the design and creation of novel hybrid materials present exciting possibilities. The incorporation of specific elements or functionalities into existing materials can significantly enhance their performance. Imagine combining organic molecules with inorganic materials to form hybrid systems that exhibit unique spin interactions and increased coherence. Such systems could lead to improved sensors for quantum sensing applications, allowing for measurements of unprecedented sensitivity and precision. Or we could create hybrid materials that are simultaneously strong, lightweight, and biocompatible, expanding the possibilities of medical imaging techniques, enhancing diagnostic capabilities, and allowing for the development of new treatment modalities.

The exploration of two-dimensional (2D) materials has revolutionized various fields, and their impact on quantum spin resonance is only beginning to be understood. Materials like graphene and transition metal dichalcogenides (TMDs) offer unique opportunities because of their atomically thin structures and remarkable electronic properties. These materials exhibit high carrier mobility and strong spin-orbit coupling, leading to potential applications in spintronics and quantum information processing. The unique electronic and magnetic properties of 2D materials pave the way for novel devices. These devices could enable the precise control and manipulation of electron and nuclear spins, creating new pathways for quantum computing and quantum sensing. Further exploration in this area may lead to even more advanced materials.

The synthesis of new materials is not just about creating something new; it is about carefully controlling the synthesis process to achieve precise control over the material's properties. This requires mastering sophisticated techniques like molecular beam epitaxy (MBE) or chemical vapor deposition (CVD), allowing for precise control over layer thickness and composition in 2D materials, or even atomic layer deposition (ALD) to fabricate complex heterostructures. These techniques provide extremely high precision in fabricating advanced materials structures. These precise techniques are crucial for ensuring that the material's properties meet the stringent requirements for quantum spin resonance applications. The more control we have over the synthesis, the better we can tailor the properties of the material to match the specific needs of the application.

Finally, theoretical modeling and computational simulations are essential partners in the quest for new materials. Computational approaches help us predict the electronic and magnetic properties of hypothetical materials before even attempting to synthesize them. This is both a huge cost saver and a powerful tool to guide the experimental efforts and prioritize promising material candidates. Sophisticated quantum mechanical calculations can provide insights into the interactions between spins in different materials and allow us to understand the factors that limit spin coherence. The combination of computational predictions and experimental results establishes a powerful feedback loop that accelerates the discovery and development of new materials, ensuring that this research is as efficient as possible.

7.4 Theoretical Developments

The theoretical landscape of quantum spin resonance is constantly evolving, driven by the quest for a deeper understanding of the underlying physics and the desire to unlock new applications. Recent advances in density functional theory (DFT) have significantly improved our ability to model the electronic structure of complex molecules and materials, paving the way for more accurate predictions of NMR and EPR parameters. This allows researchers to connect experimental observations with the microscopic details of the systems under investigation, leading to a more complete and insightful interpretation of experimental data. For example, DFT calculations can predict the hyperfine coupling constants in EPR spectroscopy with remarkable precision, providing valuable information about the electronic structure surrounding the paramagnetic center.

Progress in many-body quantum mechanics is crucial. Developing sophisticated theoretical frameworks to account for the intricate interactions between numerous spins within a sample remains a challenging but vital area. These methods often involve complex numerical simulations, like quantum Monte Carlo methods, to overcome the difficulties inherent in solving the many-body Schrödinger equation. The insights gained from these simulations offer a pathway to understanding collective phenomena, such as spin diffusion and spin-lattice relaxation, which are critical for interpreting the dynamics observed in NMR and EPR experiments. Furthermore, these theoretical models can help predict the behavior of novel quantum materials, opening up exciting possibilities for future technological applications.

The investigation of topological quantum phenomena in spin resonance is another exciting area of development. Topological insulators and other novel materials exhibit unique electronic properties linked to their topological order, profoundly affecting their spin dynamics. Understanding the interplay between topology and spin resonance requires sophisticated theoretical tools, including topological quantum field theory, to explain exotic phenomena like topological spin textures and edge states. These theoretical developments are not only academically enriching but also have significant implications for the development of spintronic devices and quantum computers, promising advanced functionalities and enhanced performance.

Theoretical exploration of quantum control and manipulation of spins is of paramount importance for advancing quantum technologies. This field is highly interdisciplinary, drawing from quantum optics, quantum information theory, and condensed matter physics. The development of advanced control techniques requires sophisticated theoretical tools and sophisticated algorithms to optimize pulse sequences, leading to improved precision and efficiency in quantum sensing, quantum computing, and other applications. Novel theoretical approaches are also being developed to characterize and quantify the quantum coherence properties of spin systems, providing a deeper understanding of the limitations and possibilities of quantum control.

Furthermore, the theoretical understanding of open quantum systems is critical. Real-world spin systems are rarely isolated; they interact with their environment, leading to decoherence and relaxation. Developing accurate theoretical models that incorporate the effects of the environment is crucial for predicting the performance of quantum technologies and devising strategies to mitigate decoherence. These models often involve master equations, quantum stochastic calculus, and other advanced mathematical techniques, allowing researchers to better understand the interplay between the system and the environment. This understanding helps in the development of new methods for extending coherence times and controlling quantum phenomena more effectively. This remains a critical challenge, as even small environmental interactions can significantly affect the performance of quantum technologies. Addressing these theoretical issues is essential to advance the field and realize the potential of quantum spin-based technologies.

7.5 Technological advancements

The relentless march of technology continues to reshape the landscape of quantum spin resonance, promising unprecedented capabilities. Miniaturization is a key driver, leading to the development of ever-smaller, more portable, and cost-effective NMR and EPR spectrometers. These advancements democratize access to these powerful techniques, enabling their application in diverse fields previously limited by the size and expense of traditional instrumentation. Imagine hand-held devices capable of performing rapid, on-site analyses, revolutionizing fields ranging from medical diagnostics to environmental monitoring.

Such miniaturization wouldn't be possible without parallel advancements in microfabrication and nanotechnology. The precision engineering required to create the minuscule coils, detectors, and other components essential for these compact spectrometers relies heavily on these cutting-edge techniques. Progress in materials science plays a crucial role, too. Researchers are constantly searching for new materials with enhanced properties—higher sensitivity, improved signal-to-noise ratios, and greater stability—leading to more efficient and robust instruments.

Furthermore, advancements in computational power directly impact the field. Analyzing the complex datasets generated by NMR and EPR experiments demands significant computational resources. The development of faster, more powerful algorithms and improved computing hardware, from powerful desktop workstations to cloud-based computing, enables scientists to extract more meaningful insights from their data more efficiently. Sophisticated software packages, featuring intuitive interfaces and advanced analytical tools, are continually refined, streamlining the data processing and interpretation phases. This is especially crucial with the increased resolution and dimensionality of modern NMR and EPR experiments.

Beyond hardware and software, improvements in pulse sequences and data acquisition methods dramatically enhance the capabilities of NMR and EPR. More sophisticated pulse sequences are continuously being developed to optimize experiments, improving signal quality and resolution. These advancements lead to better sensitivity and more specific information about the quantum systems under study. New acquisition techniques enable faster data collection with higher accuracy and efficiency.

The integration of artificial intelligence and machine learning algorithms is transforming the analysis of spectral data. These algorithms can identify subtle patterns and correlations in complex NMR and EPR spectra that might be missed by human analysis, offering unprecedented levels of information extraction. AI-powered tools help improve signal processing and noise reduction, further enhancing sensitivity and resolution. They also expedite data interpretation, accelerating the research cycle and enabling researchers to focus on the scientific interpretation rather than laborious data processing.

Finally, the convergence of quantum spin resonance with other cutting-edge technologies presents exciting possibilities. The integration of NMR and EPR with other spectroscopic techniques, like optical spectroscopy, offers synergistic approaches to studying complex systems. Combining these techniques often provides a more complete picture of the properties and dynamics of the quantum system under examination. These advancements create powerful new research tools capable of addressing increasingly complex scientific problems. The future of quantum spin resonance promises to be particularly bright, driven by the continuous push towards innovation in all related areas. The collaborative synergy between the different technological advancements further ensures an exciting future for these powerful spectroscopic techniques. New horizons continually emerge, shaping the landscape of research and applications across multiple disciplines.

www.ingramcontent.com/pod-product-compliance
Lightning Source LLC
LaVergne TN
LVHW022014200926
848481LV00007B/2391
* 9 7 9 8 3 4 9 2 8 9 6 5 1 *